Polynomials, Dynamics, and Choice

Working out solutions to polynomial equations is a mathematical problem that dates from antiquity. Galois developed a theory in which the obstacle to solving a polynomial equation is an associated collection of symmetries. Obtaining a root requires "breaking" that symmetry. When the degree of an equation is at least five, Galois Theory established that there is no formula for the solutions like those found in lower degree cases. However, this negative result doesn't mean that the practice of equation-solving ends. In a recent breakthrough, Doyle and McMullen devised a solution to the fifth-degree equation that uses geometry, algebra, and dynamics to exploit icosahedral symmetry.

Polynomials, Dynamics, and Choice: The Price We Pay for Symmetry is organized in two parts, the first of which develops an account of polynomial symmetry that relies on considerations of algebra and geometry. The second explores beyond polynomials to spaces consisting of choices ranging from mundane decisions to evolutionary algorithms that search for optimal outcomes. The two algorithms in Part I provide frameworks that capture structural issues that can arise in deliberative settings. While decision-making has been approached in mathematical terms, the novelty here is in the use of equation-solving algorithms to illuminate such problems.

Features

- Treats the topic—familiar to many—of solving polynomial equations in a way that's dramatically different from what they saw in school
- Accessible to a general audience with limited mathematical background
- Abundant diagrams and graphics.

Scott Crass is a professor of mathematics at California State University, Long Beach, where he created the Long Beach Project in Geometry and Symmetry. The project's centerpiece is The Geometry Studio, where students explore math in experimental and perceptual ways. Advised by Peter Doyle, his Ph.D. thesis at UCSD was *Solving the Sextic by Iteration: A Complex Dynamical Approach*. His research interests involve blending the algebra and geometry induced by finite group actions on complex spaces, in an effort to discover and study symmetrical structures and associated dynamical systems. A prominent feature of his work involves using maps with symmetry in order to construct elegant algorithms that home in on a polynomial's roots.

Polynomials, Dynamics, and Choice
The Price We Pay for Symmetry

Scott Crass
California State University
Long Beach, United States

CRC Press
Taylor & Francis Group
Boca Raton London New York

CRC Press is an imprint of the
Taylor & Francis Group, an **informa** business
A CHAPMAN & HALL BOOK

First edition published 2023

by CRC Press
6000 Broken Sound Parkway NW, Suite 300, Boca Raton, FL 33487-2742

and by CRC Press
4 Park Square, Milton Park, Abingdon, Oxon, OX14 4RN

CRC Press is an imprint of Taylor & Francis Group, LLC

ISBN: 978-0-367-56520-6 (hbk)
ISBN: 978-0-367-56493-3 (pbk)
ISBN: 978-1-003-09816-4 (ebk)

DOI: 10.1201/9781003098164

Typeset in CMR10 font
by KnowledgeWorks Global Ltd.

Publisher's note: This book has been prepared from camera-ready copy provided by the authors.

For Bourg, Bumper, and Monkey

Contents

List of Figures

List of Tables

Preface: Motivation

FORK IN THE ROAD

> I shall be telling this with a sigh
> Somewhere ages and ages hence:
> Two roads diverged in a wood, and I–
> I took the one less traveled by,
> And that has made all the difference.

–Robert Frost, *The Road Not Taken*

What properties characterize a human? The question reaches back at least to Aristotle and remains disputable today. In spite of the controversy, few would deny that a core human trait is the capacity to make choices. We decide one thing or another frequently and, in many cases, automatically. Indeed, the act of choosing has an inescapable quality; refraining from making a choice is itself an exercise of choice. A difficult problem for philosophy as well as for neuroscience turns on whether our choices are really choices—actions that we undertake voluntarily. Rather than take a position in that debate, let's agree to a less contentious opinion, namely, that we *feel free* when selecting among options. When we choose, we have the impression that we could have done otherwise.

Sensing that we can choose freely allows for the possibility that some choices are tough to make. What conditions lead to such a result? Does it matter how consequential a decision might be? Is it difficult to decide to pass at a blind curve on a lightly-traveled road? On a heavily-traveled road? How should we think about situations like this? Compare the pass/no-pass decision to choosing a throw in a game of rock-paper-scissors. The cases seem to be different in some basic way. But, which decision is more difficult? According to the account that we'll develop, the rock-paper-scissors choice is the harder of the two, in spite of its relative insignificance.

Among the variety of choices that we can face, the most elemental is one that's *binary*—when there are two options. Decisions of this sort are familiar: whether or not to accept a job offer, to purchase one brand of a good over another, to cast a vote for one of two candidates, to dine in a particular restaurant rather than another. Consider the binary choice before the poet quoted above. Although it surely diminishes the poem's force, let's interpret the setting literally. Is the decision facing the poet a challenging one? Would it increase or decrease the difficulty were the roads equally traveled?

A fable, ancient in origin and nominally attached to the medieval philosopher Jean Buridan, features a hungry mule in a perplexing state: two identical piles of hay are placed equidistant from the mule's head. Unable to choose between piles, the mule starves. Just what lesson is to be drawn from the fable has been discussed for more than two millennia. We'll refrain from entering the philosophical fray and will instead strive to capture the grounds of the mule's difficulty.

In a variant of the story, apparently proposed by Aristotle, one pile is hay and the other is something else that also appeals to the mule—oats, say. The political cartoon in Figure 1 alludes to this type of choice as an editorial comment on the options before the United States congress regarding the location of a canal across Central America.

Perhaps much of history's unfolding comes down to a chain of choices like the ones discussed above. To name a very few around the second world war: Germany's invasion of Poland, Japan's attack on Pearl Harbor, the D-Day invasion, and the United States's development and use of a nuclear weapon. A more recent case that Stephen Johnson examines in considerable detail is the Obama administration's decision to attack a compound in Pakistan that they had reason to believe was housing Osama bin Laden [25]. Of course, we're tempted to speculate about how things would have turned out had a different course been taken. What if you had selected a different college, a different major, or taken a different job? Which choices were made with difficulty? What's the source of the predicament? Which ones were momentous, in the sense that it's likely that the current of your life would have run in sharp contrast to the way things actually worked out?

A key question concerns how the two versions of Buridan's mule compare. What mechanism leads to the animal's inability to select one pile over the other in each case? Are the mechanisms the same? Were one of the piles to consist of straw, something the mule finds distasteful, the decision is easy. This work's first part will develop frameworks that can be brought to bear on problems like the one confronting Buridan's mule as well as many others. 'Many' refers to an infinite number of cases for which there are infinitely many options.

Whether or not you go in for Dostoevskian psychological determinism when choosing a course of action, the means by which we make such decisions is worth subjecting to inquiry. Our treatment of the conditions under which we find some choices to be difficult to make will draw on a classic problem in theoretical mathematics: working out the solutions to a specific type of equation. Part I describes and contrasts two procedures that manufacture solutions to *polynomial equations*. At the second method's core, we find elegant geometric structures and special dynamical processes that pair with the equations that we want to solve. The glue that binds an equation to geometry and dynamics is the concept of symmetry—a profound idea that plays a powerful role throughout math and science. A foundational discussion will lay out a mathe-

Figure 1 A political version of Buridan's mule. (William Allen Rogers, circa 1900, *New York Herald*. Public domain, via Wikimedia Commons).

matical formulation of symmetry as well as the process of symmetry-breaking, a crucial mechanism in the equation-solving machinery.

In the book's second part, we apply the mathematical characterization of symmetry from Part I to a more general problem of selecting among alternatives. With this analytical tool, we can apply algorithmic methods to choices like the one before Buridan's mule as well as the decisions, both consequential and not so much, that are the stuff of daily life.

METHOD OF DISCOVERY

Most of the mathematical results discussed in Chapter 5 were found by the author using both theoretical and computational tools. Standard practice involved using theory to guide computation. There was also a role for pursuing hunches that certain structures or objects might exist. Numerical and graphical experiments would then confirm or not the veracity of these conjectures. Technical details appear in the collection of referenced articles. *Mathematica* code that derives and implements results is available at [12].

OBJECTIVES

In this book's first part we explore properties and structures associated with polynomials. On its face, a polynomial is an object of pure algebra. However, this work's core idea of symmetry provides for a geometric interpretation of what it takes to solve a polynomial equation. Some of the development involves technical algebra at a high school level. Technicalities are included to the extent that they convey something of the nuts and bolts out of which a method for solving equations can be built. Grasping the essential features of the process does not require a mastery of intricate details. Much of the exposition occurs through visual means in an abundance of diagrams and graphical data that depict structures in both two and three dimensions. Readers can expect to attain an understanding of equation-solving in which symmetry weaves together conceptual threads in algebra, geometry, and dynamics.

In Part II we take a journey through assorted spaces of choices with a field guide inspired by the mathematical account found in the first part. Along the way, we encounter a variety of settings that require a decision of some kind. These deliberative situations range from the commonplace to difficult issues pertaining to optimization. Discussions are mostly non-technical or only partially so. By overlaying an algorithmic framework gleaned from equation-solving, the goal is to promote a deeper awareness of how to go about making certain choices.

Acknowledgments

The seed that grew into this work was planted during my second year of grad school at UCSD when I met Peter Doyle and expressed an interest in finding a thesis topic. Using dynamics on complex numbers, he and Curt McMullen had recently worked out a solution to the quintic equation. He suggested that I have a go at developing a similar result for the sixth-degree equation. A thesis led to a program of study that explored a number of other cases. This undertaking has been deeply enjoyable, as it draws on ideas and theory in algebra, geometry, and dynamics. At bottom, the project is aesthetic—a search for elegance.

After many days spent in a Urey Hall computer lab, I managed to solve the sextic and finish grad school. Since then, over a number of years, I've had the delight of continuing to benefit from Peter's deep insights. His influence imbues the research discussed in these pages. Saying thanks is hardly adequate, but will have to suffice.

My experience at UCSD was seasoned with friendships as well. There are fond memories of coffee at the Grove, lunch at Ché café, pints at the pub, an ascent of Mount Whitney, surfing struggles, and grad student softball. I often think of the friends with whom I shared those times: Ian Agol, Hugh Howards, Slava Krushkal, Mike Leonard, and Anne Shepler. They all have something to do with this book's existence. Peter predicted that I would become nostalgic for grad school after leaving. How right he was.

I

Polynomials: Symmetries and Solutions

CHAPTER 1

Solving Equations: A Fundamental Problem

Equations are the lifeblood of much of science and mathematics. Throughout the body of a theory, they carry material that expresses relationships among ideas and quantities that range from concrete to abstract. They can express laws that capture the behavior of physical processes: from motion ($F = ma$ and $E = mc^2$) to electromagetic interactions (Maxwell) to geometric properties of spacetime (Einstein) to fluctuations in quantum fields (Schrödinger). Of course, equations play a crucial role in the study of many other phenomena such as chemical reactions, protein structure, economic behavior, and disease transmission.

In these cases, equations constitute mathematical models, meaning that they describe relations between certain properties of a system. Models of this sort can take the form

$$f(x) = 0.$$

Here, f is a function of the variable x each of which can take values in any dimension—even infinite.

1.1 POLYNOMIAL PRIMER

Our focus will be the most elementary type of function—and the most ancient. Dating back at least as far as Babylon, humans have recognized the significance of a certain type of equation. Think of a Babylonian farmer who wants to plant a field in the shape of a rectangle with a perimeter of 50 cubits and enclosed area of 400 square-cubits. One way the farmer could try to determine the length and width of the field that solves the problem is by guessing values. If the numerical values aren't especially favorable, this method proves to be quite difficult. A more sophisticated approach represents the situation with a model. To that end, we'll use algebra—a technique that derives its power from assuming at the outset that we already possess the solution.[1] The crucial step is the first: take L and W to be the length and width of the field as Figure 1.1 indicates. State the two conditions:

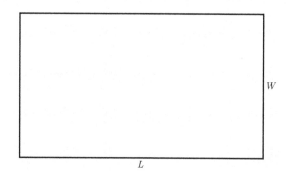

Figure 1.1 Labeled rectangle for the Babylonian farmer's problem. (Unless otherwise indicated, *Mathematica* is responsible for displayed graphics.)

$$\text{perimeter:} \quad 2\,L + 2\,W = 100$$
$$\text{area:} \quad LW = 400.$$

Next, use the perimeter equation to express W in terms of L:

$$W = 50 - L.$$

Then substitute for W in the area equation to obtain

$$50\,L - L^2 = L(50 - L) = 400,$$

which we can rearrange as

$$L^2 - 50\,L + 400 = 0.$$

If we follow common usage by taking x as a variable quantity, the expression on the left side has the form of a *polynomial*

$$a_2\,x^2 + a_1\,x + a_0.$$

The *coefficients* a_0, a_1, a_2 stand for numerical values that we can select freely; but once selected, they remain constant. We call the value of the largest exponent of the variable that appears the polynomial's *degree*. An important issue concerns the collection of numbers to which the coefficients and variable

---

[1]The treatment here is anachronistic. Babylonian mathematics did not make use of symbolic variables. Rather, a verbal description of the model was used.

belong, one that we'll examine in due course. For now, we won't specify from where the values come.

The first thing to notice is that there are polynomials of any degree:

$$a_n x^n + \cdots + a_0$$

and we typically assume that a_n is not zero. So, the simplest expression is the *zero polynomial* when all of the coefficients are zero. Next on the ladder of simplicity are the constants—when all coefficients equal zero except a_0. For the simplest non-constant polynomial, we have the first-degree family

$$a_1 x + a_0 \qquad a_1 \neq 0.$$

By substituting values for x, a polynomial becomes a function

$$P(x) = a_n x^n + \cdots + a_0.$$

Extending the degree to infinity takes us from the realm of polynomials to the that of power series—a deep topic in calculus. A remarkable feature of polynomial functions that's related to their association with power series is their capacity to approximate many non-polynomial functions to arbitrarily high precision. In the theory of computation, polynomial time, and space play a central role. Due to formidable properties such as these, polynomials became a classic field of study. They are part of the school math curriculum the world over. Of particular interest and the core concern of the first part of this book is the problem of solving a *polynomial equation* (or just equation when the context is clear)

$$P(x) = 0.$$

By solution we mean a value r, called a *root*, that satisfies the equation when r replaces x:

$$P(r) = 0.$$

Searching for ways to determine the roots of a polynomial is an elemental quest that touches upon many developments in mathematics.

The first method of solution that a student learns involves the important technique of factoring, whereby a polynomial $P(x)$ is split into simpler pieces that are themselves polynomials. The new expressions are simpler in that their degrees are lower than that of $P(x)$. Take the Babylonian farmer's equation, for example. The polynomial can be split into two factors and the equation takes the form

$$(L - 10)(L - 40) = L^2 - 50\,L + 400 = 0.$$

What this simplification tells us is that the product of the two degree-one factors on the left is equal to zero. Since the only way that can happen is for one of the factors to be zero, the roots are 10 and 40. Note that if $L = 10$

or $L = 40$, the defining equations of the farmer's problem require $W = 40$ or $W = 10$ respectively.

Searching for factors is an effective strategy only when the polynomial has a special form and, as a method, increases in difficulty with the degree. Nevertheless, factoring serves a theoretical purpose. First, the coefficient a_n of a degree-n polynomial $P(x)$ doesn't equal zero, by definition. So, we can divide $P(x)$ by a_n to get a new polynomial P/a_n whose x^n coefficient is one. It's clear that the solutions of $P = 0$ and $P/a_n = 0$ are exactly the same.[2] Since our goal is to determine roots, we can take a degree-n polynomial to have the slightly simpler form

$$P(x) = x^n + a_{n-1} x^{n-1} + \cdots + a_1 x + a_0.$$

Now, suppose P has a root—call it z_1. According to a key property of polynomials, P can be factored as

$$P(x) = (x - z_1)P_1(x).$$

The degree of P_1 is $n - 1$ which is one less than P's degree. The polynomial $x - z_1$ accounts for the missing degree. If we apply the same process with z_2 as a root of P_1, the result is

$$P(x) = (x - z_1)(x - z_2)P_2(x)$$

where P_2 has degree $n - 2$. If we can continue finding roots of the new polynomials P_2, P_3, and so on, the procedure has to end when the degree reaches zero. At that point, P has been thoroughly factored:

$$P(x) = (x - z_1)(x - z_2) \cdots (x - z_n).$$

From here, we can see that a degree-n polynomial can have no more than n roots. But, can it have *fewer* than n roots?

A quick answer to the query seems to be yes in light of

$$x^2 + 2x + 1 = (x + 1)^2$$

which has only one root, namely -1. But, this claim fails to count the roots properly. We should assign to -1 a *multiplicity* of two, meaning that it acts like two distinct roots. In the factorization of P, we permit the roots to be repeated and so, count them accordingly.

1.2   WHAT NUMBERS DO WE USE?

A more serious issue arises when we take into account the set of numbers from which the variable and coefficients are drawn. Say, we allow the use of only

---

[2] For convenience and clarity, we sometimes don't explicitly mention the variable—x in this case.

integer values—whole numbers $0, \pm 1, \pm 2$, and so on. Under this restriction, we can't solve

$$2x - 3 = 0$$

since the solution is $x = \frac{3}{2}$ which is not an integer. We can deal with this difficulty by enlarging the set of what we call *available numbers*[3] to include "rationals"—ratios of integers. But now, the solution to

$$x^2 - 2 = 0,$$

which is $x = \sqrt{2}$ (and $x = -\sqrt{2}$), is not available. This failing is due to the fact—known to antiquity—that the square-root of 2 is *irrational*, meaning that it's not equal to a ratio of integers. By the way, why should we think of $\sqrt{2}$ as a number? One justification is to observe, as Figure 1.2 illustrates, that it's the length of the diagonal of a square whose sides have length 1. Now, to remedy the shortcoming, we add both rationals and irrationals to the set of available numbers. The result of this enlargement is the *real number line* where each point on a line corresponds to a real number by virtue of the length and direction of the segment formed by the point and a reference location that we designate as 0.

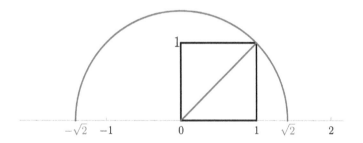

Figure 1.2 Line of real numbers. The arc is a half-circle centered at 0 and $\pm\sqrt{2}$ appear as points on the line.

Since we include all lengths, the number line is full of points; there are no gaps between any pair of numbers. You might think that nothing more needs to be added to the set of available numbers. So, using only real numbers for coefficients and variables, can we solve every polynomial equation? The case of

$$x^2 + 1 = 0$$

belies the claim that we can. When you square a real number—multiply it by itself—the result is a positive number, even when that number is negative. As the rule learned in school says: negative times negative is positive.[4]

---

[3]The terminology is not standard, but is convenient for our purposes.

[4]This rule is the subject of a math joke.

Alice: You hear that a negative times a negative is a positive, but you never hear that a positive times a positive is a negative.

Bob: Yeah, Yeah.

To solve the equation here you need a number whose square is -1, a property that no real number possesses. In fact, there are infinitely many equations with real coefficients for which no solution exists among real numbers. Following precedent, we can augment the set of available numbers to include such solutions. Take the polynomial $x^2 + 1$. For it to have a root, the value must be a square-root of -1. Call that value $i = \sqrt{-1}$ and add it to the real numbers. But how to do that? No space remains on the number line for any new points. First, note that if we replace 1 by any positive number a, $x^2 + a$ has the root $\sqrt{-a}$ which, by applying standard algebraic rules, is equal to $\sqrt{a}\,i$. Here, we're taking the square-root of a positive real number for which there are two distinct values, one positive, and the other negative. To illustrate, the square-roots of 4 are 2 and -2, a fact that can be given succinctly as $\sqrt{4} = \pm 2$. If we apply the same reasoning to square-roots of negative numbers, we get $\sqrt{-a} = \pm\sqrt{a}\,i$. So, taking y to vary over all real numbers, we should include the quantities $y\,i$ in the set of available numbers. Use of the variable y suggests that we treat this collection of *imaginary* numbers as points on a y-axis. The term 'imaginary' was applied at a time when $\sqrt{-1}$ wasn't thought of as a standard or legitimate number and the name stuck.

With two axes in mind, we can depict a new set of available numbers as an xy-plane, as in Figure 1.3. However, rather than refer to a point in this plane as (x, y)—rectangular coordinates familiar from school geometry, we characterize it by $x + y\,i$. The meaning of '+' here is not ordinary addition; rather it's a symbol that separates the *complex* number $z = x + y\,i$ into two parts: a *real part* x and an *imaginary part* y. Be careful to note that each part consists of a real number. Just what arithmetic properties the symbol '+' has is up to

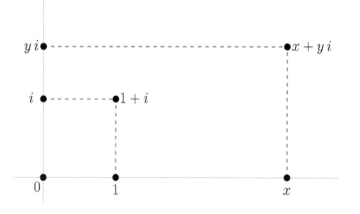

Figure 1.3 Line of complex numbers.

us. Obvious ones are: 1) for two complex numbers to be equal their real and imaginary parts must be the same and 2) to add two complex numbers add their respective real and imaginary parts. When it comes to multiplication, follow your nose by defining the product of two complex numbers as

$$(x + y\,i)(u + v\,i) := xu + x(v\,i) + (y\,i)u + (y\,i)(v\,i).$$

We can simplify this result by assuming a commutative property, namely, that a real number times i is equal to i times that real number. Applying this rule and recalling that $i^2 = -1$ gives

$$(x + y\,i)(u + v\,i) = xu + xv\,i + yu\,i + yv\,i^2$$
$$= (xu - yv) + (xv + yu)\,i.$$

When two real numbers are juxtaposed—such as xu, we mean ordinary multiplication. Much of the beauty associated with complex numbers stems from the geometric behavior of their product, though it won't be explored here. (For an extensive discussion, see [38].) We'll content ourselves with two observations of familiar behavior: 1) multiplying any complex number by zero returns zero and 2) the product of two complex numbers is itself a complex number.

In the figure, the set of complex numbers is called a line (many sources call it a plane). A rationale for such a description is the observation that, given any complex number w that's not zero, the entire set of complex numbers is realized by taking products zw as z ranges over all complex values. This sort of one-dimensional scaling is typical of a line—in this case, a complex line.

A complex number z also has a *length*, designated $|z|$, which is its distance from 0. To characterize a circle of radius R centered at 0, take all the points whose length is R as in Figure 1.4. Circles in the complex line play a significant role in later developments.

1.3 ROOTS AND COEFFICIENTS

Returning to the matter of equation-solving: Are there polynomials whose coefficients are complex numbers—call them complex polynomials, but whose roots are not? From early in the nineteenth century came the astonishing answer, now known as the *fundamental theorem of algebra*. If you search the sack containing all complex polynomials, you won't find one that fails to have a complex root. An equivalent statement of the fundamental theorem is the assertion that any degree-n complex polynomial is equal to a product of n factors each of which is a complex polynomial of the first degree. Expressed symbolically:

$$z^n + a_{n-1}\,z^{n-1} + \cdots + a_1\,z + a_0 = (z - z_1)(z - z_2)\ldots(z - z_{n-1})(z - z_n). \quad (F_n)$$

Comparing the two sides of this equation, we can see what we're up against when attempting to determine a polynomial's roots. The equation—called (F_n) for "factorization of degree-n"—establishes relationships between the

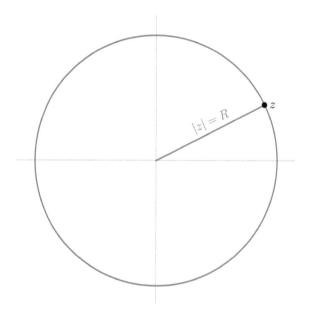

Figure 1.4 A circle in the line of complex numbers.

coefficients a_0, a_1, etc. and the roots z_1, z_2, etc. To see how this works, take the quadratic case[5] where $n = 2$ and, when multiplied out, (F_2) has the form

$$z^2 + a_1 z + a_0 = (z - z_1)(z - z_2) = z^2 + (-z_1 - z_2) z + z_1 z_2.$$

A polynomial equation is a kind of game: "I'm thinking of a number that, when you multiply it by itself, add the product of a_1 and that number, and then add a_0, you get nothing. What number(s) am I thinking of?"

By their design, two polynomials are equal exactly when their respective coefficients are the same. In equation (F_2), respective coefficients have the same color. If we set them equal to each other and move a minus sign from one side to the other, two *root equations* result:

$$z_1 + z_2 = -a_1 \quad \text{and} \quad z_1 z_2 = a_0.$$

It's worth recalling that a_0 and a_1 are *parameters*—values that are selected freely, but are constant thereafter. These equations lead to a different version of the polynomial game: "I'm thinking of two numbers whose sum is $-a_1$ and whose product is a_0. What numbers do I have in mind?"

Taking z_1 and z_2 to be variables, their sum, and product are functions. The problem here is to determine the values that z_1 and z_2 must have when you know only the values assigned to the sum and product functions. For the

---

[5]The first-degree case is trivial, since $z_1 = -a_0$.

Babylonian farmer, this leads to the original statement of the problem: find the length L and width W of a rectangle so that

$$L + W = 50 \quad \text{and} \quad LW = 400.$$

This conundrum is an instance of a classic and abstract type of mathematical difficulty called an *inverse problem*. Say you have a bunch of functions of a number of variables and each of the functions is assigned a value of the appropriate sort. (The values need not be numerical, though that's what they are in the case of polynomial equations.) Work out the values of the variables that, when substituted into the functions, produce the prescribed values. Following this kind of procedure is called inverting the functions in question. It amounts to an "undo" process of which examples abound: putting on and taking off a sock, adding a word to a document and then reverting to the original, reversing any reversible process. For an overly simple mathematical example, take the equation

$$f(x) = 3x + 1 = 2.$$

The function $f(x)$ can be undone by applying another function—the inverse of $f(x)$. In this case, it's

$$x = \frac{y - 1}{3}.$$

You can tell that this function inverts $f(x)$ by replacing the y in $\frac{y-1}{3}$ with $3x+1$ and then simplifying the expression to the original variable x. To solve the original inverse problem, all we have to do is set $f(x)$ equal to the variable y and evaluate the inverse function when $y = 2$. Doing so yields a *unique* solution:

$$x = \frac{2 - 1}{3} = \frac{1}{3}.$$

Now, you might well ask: Why make this problem so complicated? Just solve for x in the original equation. True enough, but not all inverse problems permit such a clear-cut solution. In fact, as we saw with the Babylonian problem, trying to invert the defining equations

$$L + W = 50 \quad \text{and} \quad LW = 400$$

by directly solving for L and W leads back to the farmer's polynomial, for which these are the root equations. The task calls for a novel approach, one that upcoming chapters will treat in detail.

First, we'll look more extensively into how the relationship between roots and coefficients plays out. Consider a cubic polynomial—whose degree is three. In this case, the supposition that we possess the roots z_1, z_2, and z_3 produces a third-degree (F_3) equation:

$$z^3 + a_2 z^2 + a_1 z + a_0 = (z - z_1)(z - z_2)(z - z_3)$$
$$= z^3 - (z_1 + z_2 + z_3) z^2$$
$$+ (z_1 z_2 + z_1 z_3 + z_2 z_3) z - z_1 z_2 z_3.$$

When we equate like-colored coefficients on the left and right sides, three root equations appear:

$$z_1 + z_2 + z_3 = -a_2 \qquad z_1 z_2 + z_1 z_3 + z_2 z_3 = a_1 \qquad z_1 z_2 z_3 = -a_0.$$

To solve the polynomial equation, we require three complex numbers z_1, z_2, and z_3 that satisfy all of these equations. Since this is the fundamental difficulty pertaining to the solution of a polynomial equation, let's call it the *general root problem*. Our first instinct toward solving it might be to attempt a direct inversion of the functions in the root equations; that is, express the roots explicitly in terms of the coefficients. We'll see that sometimes this works, but, in the general setting, it's a subtle and vexing question that led to profound mathematical ideas. The method at the heart of this work will solve the root problem in an indirect fashion.

From the forms of the root functions for the second and third-degree equations, we can glimpse a pattern emerging. Our treatment will, for the most part, concern polynomials of degrees two through five. Table 1.1 provides the root functions for fourth and fifth-degree cases.

Table 1.1 Root Equations for Polynomials.

Degree	Root equations
4	$z_1 + z_2 + z_3 + z_4 = -a_3$
	$z_1 z_2 + z_1 z_3 + z_1 z_4 + z_2 z_3 + z_2 z_4 + z_3 z_4 = a_2$
	$z_1 z_2 z_3 + z_1 z_2 z_4 + z_1 z_3 z_4 + z_2 z_3 z_4 = -a_1$
	$z_1 z_2 z_3 z_4 = a_0$
5	$z_1 + z_2 + z_3 + z_4 + z_5 = -a_4$
	$z_1 z_2 + z_1 z_3 + z_1 z_4 + z_1 z_5 + z_2 z_3 + z_2 z_4 + z_2 z_5 + z_3 z_4$ $+ z_3 z_5 + z_4 z_5 = a_3$
	$z_1 z_2 z_3 + z_1 z_2 z_4 + z_1 z_2 z_5 + z_1 z_3 z_4 + z_1 z_3 z_5 + z_1 z_4 z_5 + z_2 z_3 z_4$ $+ z_2 z_3 z_5 + z_2 z_4 z_5 + z_3 z_4 z_5 = -a_2$
	$z_1 z_2 z_3 z_4 + z_1 z_2 z_3 z_5 + z_1 z_2 z_4 z_5 + z_1 z_3 z_4 z_5 + z_2 z_3 z_4 z_5 = a_1$
	$z_1 z_2 z_3 z_4 z_5 = -a_0$

Knowing that inversion of the root equations is the obstacle to solving the puzzle that is a polynomial equation, we're nearly ready to put the pieces together. Actually, we examine two distinct routes to the goal, one through algebra and the other—our primary focus—through geometry. Before getting into specifics, we need to build a framework out of algebra and geometry within which a root problem can be viewed. A polynomial's connection to algebraic and geometric structure relies on the idea of symmetry—the organizing principle at this work's core. And so, we begin there.

CHAPTER 2

What Is Symmetry?

2.1 MIRRORS AND REFLECTIONS

Although 'symmetry' is a familiar term, many of us would have difficulty characterizing it precisely and succinctly. Expect to hear references to balance and pattern. Mention might be made of something having two "sides" and one side looks like the other or that the two sides are in balance. Most often, a conception of symmetry will be as a visual quality.

In school, the concept is presented in a way that's similar to how students learn colors by considering examples. So, what sort of thing has or exemplifies symmetry? When presented with several figures—usually ones that reside in a flat plane—students are told which ones are symmetric and which ones aren't. (To learn colors, replace 'symmetric' with color words.)

Figure 2.1 illustrates a learn-by-example modality. Comparing the four squares that appear, what property might we extract from the labels and placement of the red segments? (Note: the student is to assume that the shapes are squares or non-square rectangles.) In each of the three symmetric cases, the exact same shape appears on either side of the dividing line. In geometry, this sameness is called congruence, meaning that the two figures can be perfectly superimposed by moving them rigidly, that is, without any stretching or shrinking. For the square labeled "not symmetric," the two rectangles created by the red line are not congruent. The rectangle with the "symmetric" label also conforms to this interpretation. However, the rectangle with a dividing line that's a diagonal violates it; the two triangles that appear are congruent and yet the arrangement lacks symmetry.

Taking a more sophisticated approach, let's think of the red segments as mirrors, not physical ones that reflect light on one side, but as mathematical mirrors that reflect points on both sides of a line. Figure 2.2 depicts what a *reflection* does. When reflected through a line, a point called X on one side of the mirror moves to a specific location denoted $S(X)$—called the "mirror image" or just "image"—on the other side. Crucially, we're treating the reflection as a *transformation* that's applied to every point in a space—a plane, in this case. The symbol $S(X)$ tells us to apply reflection S to point X, a for-

DOI: 10.1201/9781003098164-2　　　　　　　　　　　　　　**13**

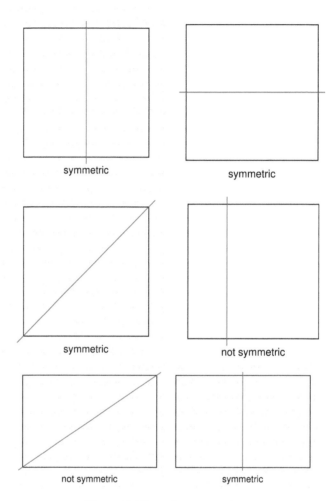

Figure 2.1 Learning symmetry.

mulation that recalls the notion of a function, which we often write as $f(x)$. As for the image point $S(X)$, it gets sent to the original point's position; we can describe this outcome succinctly:

$$S(S(X)) = X.$$

The locations of a point and its image obey a geometric rule: the line segment between the two points is perpendicular to and bisected by the mirror line. What about points *on* the mirror? Imagine that you're riding on a point that's moving closer to the mirror while the distance between the point and its image shrinks to nothing. So, it makes sense that the image of a point on the mirror is that same point.

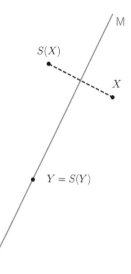

Figure 2.2 Mathematical reflection. The red line M is a mathematical mirror. Reflection through M has the effect of moving point X on one side of the mirror to its image point $S(X)$ on the other side. The mirror line forms a right angle with the dashed segment between the two points and also divides it into two pieces of equal length. Point Y resides on M and so, remains in place.

Returning to the matter of symmetry, we say that a figure is symmetric under a reflection S—or, has *reflective symmetry*—when the figure appears to be unchanged after applying S. In the examples on Figure 2.1, you can verify the correctness of the labels. To make clear what it means for the appearance of a transformed figure to be the same as the original, think of a square or rectangle as an infinite collection of points called $\mathcal{O}$ for "object." Referring to reflection through the red line as S, the reflected collection of points that results from applying S to each point in $\mathcal{O}$ is called $S(\mathcal{O})$. That the reflected object appears to be the same as the original object amounts to the collection of reflected points being the same as the original collection. Put symbolically:

$$S(\mathcal{O}) = \{\text{reflected points}\} = \{\text{original points}\} = \mathcal{O}.$$

Importantly, this does not say that each point in the figure is unmoved. Indeed, only two points are *fixed*—that is, maintain their position after reflection. Rather, what remains the same is the collection *as a whole*. Figure 2.3 displays the result of reflecting a rectangle through a mirror along a diagonal. Since the reflected rectangle is not the same set of points as the original, this arrangement lacks reflective symmetry for this particular reflection. Note that the rectangle is symmetric for two other reflective transformations.

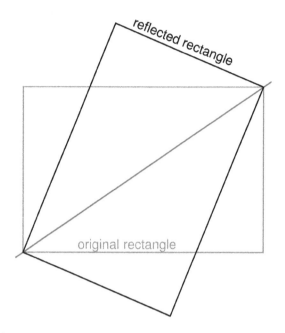

Figure 2.3 Reflecting a rectangle through a diagonal line shows that the image rectangle (black) is not the same set of points as the original rectangle (gray).

2.2   MATHEMATICAL SYMMETRY

By regarding the set of points that a figure occupies as a *property*—call it the object's location, these examples supply three ingredients that go into an abstract characterization of symmetry.

> A **symmetry** of an object $\mathcal{O}$ is a transformation T that leaves some property P of $\mathcal{O}$ unchanged.

We say that $\mathcal{O}$ has symmetry T relative to property P. As the term suggests, the notion of object here is very broad, ranging from physical and geometric forms to abstract structures and configurations. Later in the chapter, we discuss a diversity of conforming instances. We're assuming that such a transformation can be undone—that is, has an inverse transformation. With this statement, we arrive at the book's central idea, which is captured in a concise, useful, and iconic form. Let's call it the *symmetry equation*:

$$P(T(\mathcal{O})) \;=\; P(\mathcal{O})$$

$$\underset{\substack{\text{value of property } P \\ \text{for the transformed object}}}{} \;=\; \underset{\substack{\text{value of property } P \\ \text{for the original object.}}}{}$$

2.3   EXPLORING GEOMETRIC SYMMETRY

Rotations and translations are common transformations with geometric flavor. Along with reflections, they're called "rigid motions" (or just motions[1]) and can, under suitable conditions, become symmetries. (Think of points filling a sheet of metal.) Figure 2.4 portrays their behavior when applied to a plane. A rotation turns the entire space about a central point—the axis of rotation. As for translation, every point slides by a fixed length in a fixed direction.

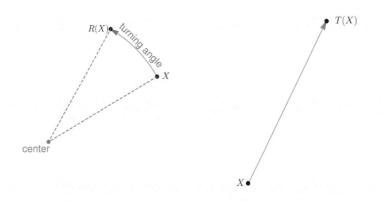

Figure 2.4 Rotation (left) and translation (right) transformations in a plane.

Using the property of location, we can recognize instances of rotational symmetry in Figure 2.5. Turning either shape by one-third of a full turn[2] (120°, if you prefer) about the marked central point returns it to the original location. Applying the rotation a second time also gives the impression that nothing has changed.

To appreciate what's happening at a deeper level, we can describe the behavior in an algebraic way. Denote by R the one-third turn, after which there's a sensible name for the repeated application of the rotation:

$$\text{(one-third turn) followed by (one-third turn)} = RR = R^2.$$

Now, R^2 is a transformation in its own right—namely, two-thirds of a full turn—that qualifies as a rotational symmetry of either figure.

The triangle is symmetric under reflections through mirrors A and B. Call these transformations A and B respectively.[3] Since the location of the triangle remains the same after each reflection, it doesn't change when applying the

---

[1] Some treatments don't consider reflections to be motions. Adopting the convention that they are motions permits a more convenient description.

[2] Following mathematical convention, a counterclockwise measurement of an angle is positive and a clockwise measurement is negative. For instance, a turn of negative one-third has the same effect on points as positive two-thirds of a turn.

[3] Note the difference in fonts.

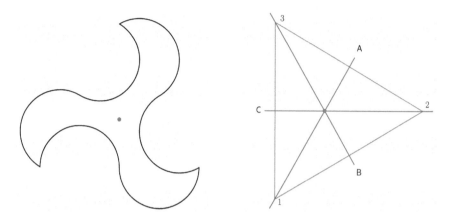

Figure 2.5 Rotational symmetry. Turning the pinwheel figure by one-third and two-thirds of a whole turn about the central point leaves its location unchanged. A reflection cannot produce that outcome. The equilateral triangle has three reflective symmetries the mirrors for which are shown. It also has the same rotational symmetries as the pinwheel, provided that the centers of each figure coincide.

compound transformation of first reflecting through A and then reflecting through B. It's natural to describe this new symmetry of the triangle as the "product" BA since it shares some properties with ordinary multiplication of numbers. Note that the transformation on the right "goes first." An important difference between products of transformations and numbers is that the order in which transformations are "multiplied" can make a difference. As we'll soon see, $BA \neq AB$.

A result of fundamental significance can be drawn from the observation that BA and AB are automatically symmetries since both A and B are.

> According to the **First Symmetry Principle**, successively applying two symmetries of an object relative to the same property produces another symmetry of that object relative to that property.

This setting is fertile ground from which the idea of symmetry grows into a mathematical theory, one that bears on the problem of solving polynomial equations. Let's see how our algebraic formulation can lead to valuable insights.

Using the symmetry equation, we can deduce Symmetry Principle I. Say we have two symmetries S and T relative to some property P of object $\mathcal{O}$. Form the compound transformation $U = TS$ (remember, first apply S and then T). Now, work through what it takes for U to be a symmetry:

$$P(U(\mathcal{O})) \overset{U=TS}{=} P(T(S(\mathcal{O}))) \overset{T \text{ is a symmetry}}{=} P(S(\mathcal{O})) \overset{S \text{ is a symmetry}}{=} P(\mathcal{O}).$$

The comment above an equal sign justifies passing from the left to the right side of the equation.

You might wonder: Is BA a symmetry that's known to us? It turns out that it has the same effect on points as does the rotation R. So, we can say

$$R = BA.$$

On exhibit here is a basic principle of transformational geometry.

- Two successive reflections through intersecting mirrors are equivalent to a rotation about the point of intersection by twice the appropriately interpreted angle between the mirrors.

- Two successive reflections through parallel mirrors are equivalent to a translation in a direction perpendicular to the mirrors by twice the distance between the mirrors.

We can check this *Two Reflections Principle* on an equilateral triangle by watching how the vertices (that is, the corners) move when two reflections are applied successively. Take note of how the labeled vertices in Figure 2.6 move when the reflections are applied—first A, then B.

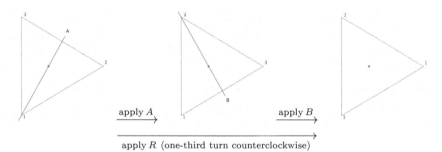

Figure 2.6 Two reflections principle. Follow the labels $1, 2, 3$ to see that successive reflections amounts to a rotation

The two reflections principle gives the result for every pair involving A, B, and C:

$$R = BA = CB = AC$$
$$R^2 = AB = BC = CA$$
$$I = A^2 = B^2 = C^2.$$

The third line conforms to a general fact: repeating a reflection leaves all points unmoved, which is what the transformation called I does. It's worth noting that by treating the triangle-with-labels as an object, each non-trivial symmetry possessed by the unlabeled triangle fails to be a symmetry of the labeled figure. The symmetries of the unlabeled triangle imply that the vertices are *indistinguishable*, a hallmark of symmetry's presence. Since labeling the vertices is a way of distinguishing them from one another, the former symmetry is now lost or *broken*. Breaking symmetry in this way can be a powerful analytic tool for quantifying and understanding an object's symmetric properties.

2.4   GROUPS IN THE ABSTRACT

By taking only the behavior of the vertex labels into consideration, the *action* of an equilateral triangle's symmetries admits an alternative description. We simply track where labeled vertices go when a symmetry is applied. A diagram is useful here—see Table 2.1. To read this notation, use the label positions

Table 2.1 Triangular Shuffles. Vertical arrows indicate where the labels associated with the vertices move relative to their reference position when the respective transformations act.

$$
I : \begin{bmatrix} 1 & 2 & 3 \\ \downarrow & \downarrow & \downarrow \\ 1 & 2 & 3 \end{bmatrix} \quad
R : \begin{bmatrix} 1 & 2 & 3 \\ \downarrow & \downarrow & \downarrow \\ 2 & 3 & 1 \end{bmatrix} \quad
R^2 : \begin{bmatrix} 1 & 2 & 3 \\ \downarrow & \downarrow & \downarrow \\ 3 & 1 & 2 \end{bmatrix}
$$

$$
A : \begin{bmatrix} 1 & 2 & 3 \\ \downarrow & \downarrow & \downarrow \\ 1 & 3 & 2 \end{bmatrix} \quad
B : \begin{bmatrix} 1 & 2 & 3 \\ \downarrow & \downarrow & \downarrow \\ 2 & 1 & 3 \end{bmatrix} \quad
C : \begin{bmatrix} 1 & 2 & 3 \\ \downarrow & \downarrow & \downarrow \\ 3 & 2 & 1 \end{bmatrix}
$$

shown in Figure 2.5 for reference and, by way of example, take the diagram associated with R—middle, top row. It records the data that when R acts on the triangle, vertex 1 moves to the position of vertex 2, vertex 2 moves to the position of vertex 3, and vertex 3 moves to the position of vertex 1. You can think of the operation represented by this sort of diagram as the shuffling of elements—in this example, the elements are 1, 2, and 3.

The first thing to address is the number of ways you can shuffle three elements. Starting with 1, there are three choices for where it can go. For each of those three possible outcomes for 1, vertex 2 can be assigned to one of the two positions that remain. In all, that's $6 = 3 \cdot 2$ possible shuffles (also known as permutations). Since no two shuffles are the same, the list given for the triangle contains all six possibilities.

The technique of counting that was employed in the case of three elements carries over to any number. Using factorial notation, Table 2.2 lists the number of shuffles for the next three cases, ones with a prominent role to play a bit later.

Just as the rotational and reflective symmetries of a triangle can combine to form another symmetry, so can we apply shuffles successively to form another shuffle. To illustrate how this goes, take the shuffles associated with rotation R and reflection B. By "stacking" the shuffles associated with these

Table 2.2 Counting Members of Complete Shuffle Groups.

# elements	# shuffles
4	$24 = 4! = 4 \cdot 3 \cdot 2$
5	$120 = 5! = 5 \cdot 4 \cdot 3 \cdot 2$
6	$720 = 6! = 6 \cdot 5 \cdot 4 \cdot 3 \cdot 2$

two transformations, we get a diagram like this:

$$A = BR : \begin{bmatrix} & 1 & 2 & 3 \\ R & \downarrow & \downarrow & \downarrow \\ & 2 & 3 & 1 \\ B & \downarrow & \downarrow & \downarrow \\ & 1 & 3 & 2 \end{bmatrix}.$$

Beginning with the top row, apply the "R-shuffle" to get the middle row. Then apply the B-shuffle to the middle row. The result is the row on the bottom, which is the effect of applying the A-shuffle to the top row. This outcome can be derived purely in algebraic terms:

$$BR = B(BA) = (BB)A = B^2 A = IA = A.$$

A transformation that does nothing might seem to be a pointless addition, as it would trivially be a symmetry for any object whatsoever. Nevertheless, including it is crucial. Why? According to the First Symmetry Principle, R followed by R^2 produces a pinwheel symmetry. But, combining these two rotations amounts to a full turn, which has the same do-nothing effect as the trivial transformation. Note that viewing I as a rotation—by zero or 360 degrees, say—is sensible. So, we must take I to be a symmetry. The upshot: you can do nothing to everything; symmetry is universal.

We can now enumerate the symmetries relative to location for the pinwheel and triangle (Table 2.3). In each case, the result is a system of transformations that can be "multiplied." The system is also *closed* in the sense that three conditions are met.

1. The trivial transformation belongs to the system.

2. The product of any pair of transformations is one of the system's transformations.

3. Each member has an inverse transformation (one that undoes it) that also resides within the system.

Put more generally, the **Second Symmetry Principle** states:

An object's symmetries relative to a specific property forms a self-contained collection of transformations called a *symmetry group*.

Table 2.3 Symmetries of a Pinwheel and Triangle.

Type	Pinwheel	Triangle
rotation	R, R^2, $R^3 = I$	R, R^2, $R^3 = I$
reflection	- -	A, B, C

To be specific, the three rotations comprise the symmetry group of the pin-wheel while the rotations and three reflections constitute the triangle's symmetry group. A complete set of symmetries is a special kind of a more abstract entity simply called a *group*. The discussion in this book will be restricted to symmetry groups, which turns out not to be restrictive. At the moment, it's not apparent that the shuffles form a symmetry group; we haven't identified a property that remains the same when each shuffle acts. The next chapter fills in that omission.

2.5   POSING AND SOLVING PROBLEMS WITH SYMMETRY

At this point a basic mathematical question arises: Does the pinwheel or triangle have more symmetries than the ones listed? Working out this problem calls for looking at the matter more carefully. A reflection is a special kind of transformation in that the distance between two points is the same as the distance between their images. Figure 2.7 depicts the situation. We should immediately observe that the distance-preserving property conforms to our characterization of symmetry. As before, if we represent the reflection through M as S and the images of points X and Y as $S(X)$ and $S(Y)$, the distances between respective pairs of points remains the same:

$$\text{distance}(X, Y) = \text{distance}(S(X), S(Y)).$$

By reverse engineering the two-reflections principle, any rotation or translation can be decomposed into a product of two reflections. It then follows from Symmetry Principle I that rotations and translations are distance-preserving as well. In fact, any transformation on the plane that doesn't distort distance can be expressed as the succession of either one, two, or three reflections.

We haven't seen an example of a distance-preserving transformation that requires three reflections—that is, one that's not equivalent to either a single reflection or a successive pair of reflections. A transformation of this sort arises as a symmetry of a certain pattern called a frieze; strips used to border wallpaper are an example. You can see a design of this kind in Figure 2.8. The pattern's symmetry is designated a *glide reflection* since it both translates along (glides) and reflects across the strip. You can simulate this behavior by marking a "P" on a transparent rectangular sheet so that the "stem" is vertical, sliding the sheet in the direction of the P's stem, and then giving the

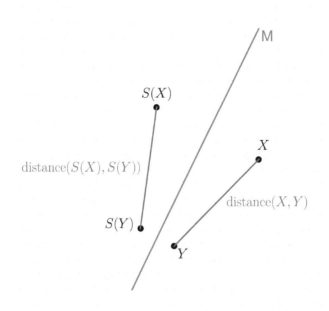

Figure 2.7 Reflection preserves distance between points. The segment between X and Y has the same length as the segment between the images $S(X)$ and $S(Y)$.

sheet a half-turn about the vertical axis. It takes both gliding and reflecting to move each P onto a P and vice-versa, an outcome that's not achievable by employing either a single reflection or, as we'll soon see, a single rotation or translation. The dots at the ends of the strip indicate the infinite continuation of the pattern. Here is another consequence of Symmetry Principle I: an object with translational or glide-reflection symmetry must have infinite extent.

Friezes can be classified by means of a criterion that declares two such patterns to be equivalent if their symmetry groups have the same structure. This condition does not require the two designs to look alike. Rather, structural similarity or difference lies in their conduct, a topic we'll take up shortly. It turns out that there are seven distinct frieze patterns.

Due to its repeating behavior, a frieze is a sort of one-dimensional crystal. Extending repeatability in a pattern to two dimensions produces crystals in a plane, as we find in wallpaper. Using symmetry groups again to classify the patterns leads to a triumph in plane geometry: there are seventeen distinct wallpaper patterns [4]. The simple definition of a crystallographic pattern applies in any dimension and their classification stands as one of the great problems of geometry.

Looking a bit more deeply, possession of a repetitive pattern's catalog of symmetries confers an ability to encode in a small piece the entirety of an infinitely extended design. In our sample frieze, the presence of glide-reflection

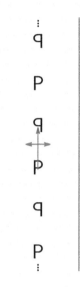

Figure 2.8 Frieze symmetry. The pattern made by P and ᑫ lands on itself when a vertical glide (blue arrow) and left-right reflection (red double-arrow) are applied sequentially.

symmetry allows for reduction of the full pattern to a single rectangle that contains just one P. This data compression relies on the capacity to generate the whole frieze by applying its symmetries to the rectangle. You can physically realize the shrinkage in a frieze's information content by carving a motif on an inked drum that creates the pattern by rolling forward and backward. In general, crystallographic structures such as wallpaper admit this sort of loss-less encoding.

Analogues to wallpaper patterns occur on the surface of a sphere. Take a soccer ball, for example, with its twelve pentagons and twenty hexagons. (See Figure 2.11.) Once again, the full structure reduces to a small part. Many viruses take economic advantage of this symmetric encoding. By virtue of the repetition driven by a pattern's symmetries, they assemble complete spherical structures by making many proteins of relatively few types. Such parsimonious construction means that a viral genome and hence a virus can be small as microbes go.

To address the question of whether our list of symmetries for the pinwheel and triangle is complete, we'll stipulate that the transformations preserve distance. Each of these objects have several types of *special points*: a one-of-a-kind center and three-of-a-kind vertices. If a transformation moves the center, it can't be a symmetry, since the transformed vertices would fail to be the same distance from both the new and original centers. So, for a rotational symmetry, the axis has to be the center while a reflective symmetry's mirror must pass through the center. In order for either figure to appear unchanged, a vertex must move to the position of a vertex—either itself or one of the other two. The only way to achieve that outcome with a rotation is by making a one-third or two-thirds turn. To move a vertex to a vertex with a reflection, exactly one of the vertices, in addition to the center, must lie on the mirror. You must then move the other two vertices so as to exchange them and thus, the mirror must bisect the edge between those two points. We can now rightfully claim that no symmetries of these figures have escaped our analysis.

An important feature of a group appears when a subset forms a group in its own right. Significantly, the number of items in such a *subgroup* is a divisor of the number of elements in the group. (For this to make sense, we assume that the group in question contains a finite number of elements.) In the case of the triangle's group, the three rotations form a subgroup, as do pairs consisting of one reflection and I. For one of the approaches to equation-solving that we'll consider, subgroups will be a crucial issue

2.6 STRUCTURE IN THE ABSTRACT

Group theory provides the framework for a mathematical account of symmetry. But, what does the group concept capture? Let's return to the association between an equilateral triangle's symmetries and the shuffles of three elements. Each system has what it takes to be a group, if we take the product of two shuffles to be the rule for combining a pair of them. However, comparing the elements of the two groups, they're quite different. Rotations and reflections act on an infinite two-dimensional space of points whereas a shuffle rearranges a three-membered set. Nevertheless, by matching the elements of each group as we have, it becomes clear that the two groups have the same structure. What's meant by the structure of a group is the way the members behave when all possible products are taken.

The abstract theory of groups seeks to classify this kind of structure using the key concept of *isomorphism*. We'll come to see that polynomial symmetry groups are isomorphic to certain other groups, a result that plays a crucial role in solving the polynomial root problem.

Maybe you wonder whether any group that consists of six elements will turn out to have the same structure as the triangle or shuffle group. Not so. Consider the rotation, call it K, by one-sixth of a turn and check that the following set of six rotations is a group:

$$\{K, K^2, K^3, K^4, K^5, K^6 = I\}.$$

This six-element group is structurally different from the triangle group. All you need to see is that, by multiplying the members, a single element generates everything in the former group, while it takes at least two elements to produce a triangle's symmetries.

2.7 A LOOK AT HIGHER DIMENSIONS

So far, our attention has been restricted to transformations on a two-dimensional space—namely, a plane. What happens in three dimensions and higher? For transformations that preserve the distance between points, the basic story, told in Table 2.4, is that the dimensions of rotation axes and mirrors increase with the dimension of the surrounding space. The relationships among these various dimensions emerge from the fact that a two-reflections principle holds in every dimension. For instance, in our familiar three-dimensional space, a mirror is a plane and successive reflections through intersecting mirror-planes acts just like a rotation about the line formed by the intersecting planes with a measure of turning equal to twice the angle formed by the mirrors. Reflecting successively through parallel planes amounts to a translation in a direction perpendicular to both planes by twice the distance that separates the mirrors.

Reasoning by analogy, we can discern the situation in four dimensions, despite our inability to visualize it completely. In this setting, a mirror is a three-dimensional space—a chunk of which is what residents of a four-dimensional world hang on a wall and peer into. When a reflection follows a reflection and the mirrors intersect, the result is a rotation about the plane formed by intersecting the two mirror-spaces. If the mirrors fail to intersect—they're parallel, you get a translation whose direction is perpendicular to both of them. When it comes to dimensions higher than four, our imagination flags. Nevertheless, geometric understanding penetrates these realms by virtue of analogical thought, algebraic description, and reduction to lower dimensions.

Table 2.4 Distance-preserving Transformations in Higher Dimensions. The middle columns show dimensions of axes and mirrors and the right-most column gives the maximum number of reflections that it takes to express a distance-preserving transformation in the given dimension.

Dimension of the surrounding space	Rotation axis	Reflection mirror	Maximum number of reflections
1	-	0D (point)	2
2	0D	1D (line)	3
3	1D	2D (plane)	4
4	2D	3D (space)	5
⋮	⋮	⋮	⋮

With the exception of distance, the instances of symmetry seen so far involve location as the unchanged property. For the remainder of this chapter, we'll take note of the breadth and depth realized by the symmetry concept.

2.8   WHAT IS GEOMETRY?

As we've seen, the distance separating a pair of points is maintained between their mirror images. Because they are combinations of reflections, the same goes for the other motions: rotations and translations. The symmetry of maintaining distance gives rise to others. In a plane, the area occupied by a figure is the same as the area taken up by that figure subjected to a motion. Recalling the symmetry equation, if T is any succession of reflections and $\mathcal{O}$ is a region that takes up area, we can say

$$\text{area}(T(\mathcal{O})) = \text{area}(\mathcal{O}).$$

The analogous consequence holds in spaces whose dimension is three or higher by replacing 'area' with 'volume' or 'hyper-volume.'

It merits a mention that Felix Klein—the intellectual forebear of the approach to polynomial equations that we will discuss at length—characterized the field of geometry in terms of symmetry groups. In his *Erlanger program*, a geometry consists of a space upon which a group of transformations acts. Having a group structure here is essential. From this point of view, what features deserve to be called geometric? We take them to be those that remain the same when any of the transformations are applied to the space. We call such a property an invariant and the set of all transformations under which this property fails to vary forms a symmetry group [30, 23]. As examples, the collection of motions in any dimension forms a group. Moreover, the properties of angle, distance, area, volume, and hyper-volume, being invariant under the relevant transformations, are thereby geometric ones. The resulting systems include the Euclidean geometries taught in school. Klein's formulation provides a powerful instrument for working with structures that arise from the action of symmetry groups, one that we use in our treatment of equations.

For the cases examined until now, we began with an object's property—location, distance—and then found transformations that leaves the property unchanged. The procedure's order is reversible; that is, starting with a group of transformations, search for a property of an object that doesn't vary when transforming the object.

Let's add *scaling* to our inventory of geometric transformations. In any dimension, to scale an object about a reference point—the center, pick any point in the surrounding space other than the center and stretch or shrink the line segment from that point to the center by a fixed factor. For a scale-factor of two, every point's distance from the center doubles. Figure 2.9 illustrates shrinking.

Suppose that we add scaling transformations to the rigid motions and include all combinations of scalings and reflections. The result is a super-group, so called because it contains the group of motions. Is there an object in the plane with a property that remains the same after all of the members of this newly-built group are applied? Figure 2.9 shows the effect of scaling on a triangle whose vertices are A, B, and C—it shrinks to a triangle with $V(A)$,

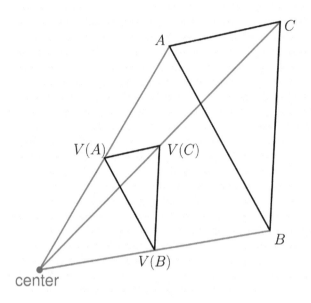

Figure 2.9 Scaling transformation. Referred to as V, its effect is to shrink by a factor of one-half. Point A is twice as far from the center as $V(A)$, the site to which it moves. The same condition holds for B and C.

$V(B)$, and $V(C)$ as vertices. Comparing the two figures, the distances separating respective vertices are not the same and thus, neither are the enclosed areas. However, the angle at a vertex of one triangle is equal to the angle at the other triangle's corresponding vertex. In other words, the angle at A equals the angle at $V(A)$ and so on. Angle measure formed by two intersecting lines is the kind of property for which we're looking. A more general interpretation is to view the two triangles as having the same shape. Indeed, the notion of *shape* can be construed as an overall geometric property of a figure that doesn't change after applying the members of the motions-plus-scalings group. You might contrast the concept of size with that of shape.

We can use assemblies of triangles to closely approximate many regions. When transforming the region, each triangle maintains its shape as does the approximated figure. Of course, such symmetry provides the rationale for building scale models of devices. We expect a scaled-down version of an airplane or bridge to behave like the full-sized entity, at least in certain important respects.

After obtaining a theoretical result, mathematicians often try to formulate a new question in light of what they've found. For instance, take the two-reflections principle and ask if an analogous result occurs for a reflection. To wit, is there a two-rotations principle stating that a reflection can be decomposed into a product of two rotations? You could approach this query by working out whether a reflection moves points around in the same way as a pair of rotations. Such an undertaking can be managed using either geometric or algebraic methods. However, we can bring symmetry considerations to bear on the problem and thereby gain valuable insight.

The idea to be exploited is known as *orientation*. Figure 2.10 illustrates this property in the plane. Looking at one of the pairs of arrows—called a "frame," define the frame's orientation by noting which way the red arrow makes a quarter turn (90°) in order to reach the position of the blue arrow. Using the convention that a counterclockwise turn is positive and a clockwise movement is negative, orientation takes the value + or −. Using the theory of vectors and matrices, orientation is definable in any dimension where a frame contains as many mutually perpendicular arrows as the dimension.

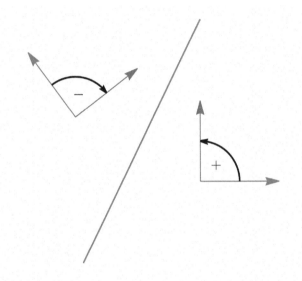

Figure 2.10 Orientation is reversed by a reflection. The mirror image of the positively-oriented (+) frame is the negatively-oriented (−) frame.

The diagram indicates how a frame and its mirror image have opposite or reversed orientations. In this phenomenon, we recognize the notion of *chirality*, a concept that plays a profound role in particle physics and biochemistry. By applying only rotations and translations *within* the plane, neither of the pictured frames can be moved onto its mirror image with the colored arrows matching. As such, they are chiral. That the orientation is unchanged when a frame undergoes a rotation or translation (scaling as well) is clear. Put another way, orientation is a symmetry-property under rotation or translation, but not reflection.

With these observations, we can resolve the matter of a two-rotations principle with ease. According to the First Symmetry Principle, any number of successive rotations produces an orientation-preserving transformation. If a reflection were equivalent to a product of rotations, it would preserve orientation, contrary to fact. As a further consequence of this line of thought, the set of orientation-preserving transformations is a symmetry group that contains all combinations of rotations and translations or, alternatively, all products of an even number of reflections.

2.9 MOLECULAR SYMMETRY

The spatial arrangement of atoms that form a molecule is critical to its chemical properties. Take the case of carbon molecules where the atoms can occupy various configurations, as depicted in Figure 2.11. In one case, each carbon atom attaches to three others in a structure that looks like a floor covered in tiles each of which has a hexagonal shape. Lying in a plane, the single-layer material that results from this system of strong bonding is graphene, a two-dimensional crystal. Stacking multiple weakly-bonded layers results in graphite. Under high pressure, distortion of graphite's carbon atoms can be made to bond to four neighbors in a structure that forms a pyramid made of triangles. The resulting "tetrahedron" requires three dimensions. Moreover, the triangular frameworks fit together to form the ultra-hard diamond crystal. In a third type of pure carbon molecule, sixty atoms arrange themselves in a soccer ball pattern called fullerene. Each particle makes single bonds to two others and a double bond to one other. Amazingly, the structure that a soccer ball exemplifies appears on stage in the equation-solving drama to come.

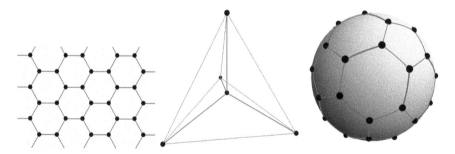

Figure 2.11 Pure carbon structures. Atoms appear as black spheres and bonds are segments in red and blue. (Left to right) Two-dimensional graphene crystal; three-dimensional diamond crystal (the gray segments outline the pyramid that forms); spherical fullerene carbon-60 with two single bonds (red) and one double bond (blue) at each atom.

Chemistry is a quest to discover and analyze what happens when molecules are made to react. The reaction process can be divided into three phases: an input batch of atoms clustered into molecules, a sequence of reactions when the material is acted upon in some way, and an output state consisting of molecules that differ from those at the outset. For the example illustrated in the diagram below, one methane and one water molecule react to produce one carbon monoxide and three molecules of hydrogen gas. We can bring our formulation of symmetry to bear on this situation with a question: When passing from the input state to that of the output, is there a property of the system that remains the same? We're viewing the reaction stage as a kind of transformation and so, our query conforms to the symmetry equation. A basic chemical principle demands equality between the number of reactant atoms (input) and the number of product atoms (output) of each kind. For the example, the reactants and products each contain one carbon, six hydrogen, and one oxygen. The symmetry property here is a *conserved quantity*: the number of constituent atoms of a specific kind. In a more general context, symmetry is expressed as a *conservation law*, one of the fundamental pieces of physical theory.

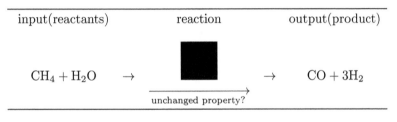

input(reactants)	reaction	output(product)

$$CH_4 + H_2O \quad \rightarrow \qquad\qquad \rightarrow \quad CO + 3H_2$$

$$\xrightarrow{\hspace{3cm}}$$
unchanged property?

2.10   CONSERVATION LAWS

One of the most basic types of physical interaction is a collision, which we imagine takes place between two balls rolling along a horizontal track without either one's shape being deformed—a so-called elastic interaction. As shown in Figure 2.12, designate the balls' masses as m and n and their velocities as u and v, respectively. The collision system conforms to an input-interaction-output framework, leaving us to search again for a property in the guise of a quantity that refuses to vary as the system evolves from input to output states. An obvious candidate is the mass of either ball, a value that we expect to be the same after collision as before.

A more sophisticated approach takes account of both mass and velocity, a task that can be simply done by forming the product of the two. The resulting quantity $p = mv$ is a particle's momentum[4] while the total momentum of a system is the sum of the momenta of all its parts. Note that velocity conveys directional information. For a ball on the track, motion to the right is positive velocity and to the left is negative. The phenomenon in question is ruled by Newton's laws of motion.[5] From the three Newtonian axioms, you can derive the conservation of momentum: so long as no force acts on it, a system's total

---

[4]Technically, it's called linear momentum, but we won't need to make the distinction.
[5]Assuming that the velocities are relatively small compared to that of light and the particles are relatively large.

momentum remains the same as its constituent parts interact. In other words, a system's input momentum equals its output momentum. For the colliding balls, this condition amounts to saying

$$mu_{in} + nv_{in} = p_{in} = p_{out} = mu_{out} + nv_{out}.$$

In more abstract terms, calling a mechanical system $\mathcal{S}$ and an interaction process T, momentum conservation can be framed as a symmetry equation:

$$p(T(\mathcal{S})) = p(\mathcal{S}).$$

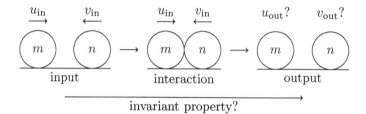

Figure 2.12 Input-interaction-output collision process. The symmetry question asks for an invariant property—a conserved quantity—when passing from the input state to the output state. Such properties can provide for quantitative results regarding the output velocities of the masses.

Of course, there's a physical problem here. If the masses and input velocities are given, what's the output velocity of each ball? Does momentum conservation tell us something in that regard? To make the example a bit simpler, let the two masses be the same and the input velocities equal in magnitude and opposite in direction:

$$m = n \qquad v_{in} = -u_{in}.$$

As for the input momentum:

$$p_{in} = mu_{in} + mv_{in} = m(u_{in} - u_{in}) = 0.$$

By conservation, the output momentum is also zero:

$$m(u_{out} + v_{out}) = p_{out} = 0$$

which means that
$$v_{out} = -u_{out}.$$

So, as a consequence of momentum conservation, the output velocities have the same magnitude and opposite directions.

However, the condition just derived fails to yield actual values for the unknown velocities. As far as we know, both output speeds could be zero—the balls collide and stick. A remedy for this state of affairs is due to a second conserved quantity: energy. Specifically, for the colliding balls, all of the energy is kinetic—that is, a property of motion. A body moving at a velocity v has a kinetic energy $K = \frac{1}{2}mv^2$. By squaring the velocity, its directional quality disappears.[6] Bringing both momentum and energy conservation to bear on our "two-body" problem produces this:

$$\frac{1}{2}m\left(u_{\text{in}}^2 + v_{\text{in}}^2\right) = K_{\text{in}} = K_{\text{out}} = \frac{1}{2}m\left(u_{\text{out}}^2 + v_{\text{out}}^2\right)$$
$$u_{\text{in}}^2 + (-u_{\text{in}})^2 = u_{\text{out}}^2 + (-u_{\text{out}})^2$$
$$u_{\text{in}}^2 = u_{\text{out}}^2.$$

From the third line it follows that either

$$u_{\text{out}} = u_{\text{in}} \quad \text{and} \quad v_{\text{out}} = -u_{\text{in}}$$

or

$$u_{\text{out}} = -u_{\text{in}} \quad \text{and} \quad v_{\text{out}} = u_{\text{in}}.$$

We can rule out the former solution as not physical, where the balls pass through one another. The upshot is that the bodies recoil at the same speed as they had prior to collision.

These and other conservation principles are consequences of a deeper symmetry that physical laws themselves respect. In the collision setting, the relevant laws are Newton's axioms that govern mechanical processes. A profound theorem discovered by Emmy Nöther in 1915 establishes a correspondence between certain symmetries enjoyed by the laws governing a system's behavior and certain quantities that are conserved as the system evolves [39]. In the context of matter in motion, conservation of momentum or energy follows from the invariance of mechanical laws when a system is translated respectively in space or time. (Translation in time can be regarded as a displacement along a time axis.)

2.11 THERMODYNAMIC SYSTEMS

Instead of having the simplest case of just two objects that interact, how can we understand a system consisting of many interacting particles—a gas in a container, say. Trying to analyze such an ensemble as we did for two bodies would lead to a morass of equations the solution to which would prove to be a quixotic pursuit. A more tractable mode of treatment is attainable through probability. While we lack the wherewithal to track the behavior of each individual particle, a coarse-grained account that treats the collection as

---

[6]In higher dimensions, the treatment uses fancier machinery.

a whole conveys insight. So, we look for large-scale properties that are realized by a many-body system.

Although the kinetic energy of an individual particle is inaccessible, the average energy over the population manifests itself in the form of temperature. In a container of gas at equilibrium with its surroundings, the hand of symmetry is evident in the system's unchanging temperature despite an enormous number of interactions taking place.

We turn next to a more subtle idea by distinguishing two levels at which the system can be described: microstate and macrostate. The former examines properties of individual particles whereas the latter concerns itself with how the population behaves. Let's use an idealized toy model to explain the theory. Illustrated in Figure 2.13 are 100 particles within a container. In the system's initial state, all of the particles are on the left side of the chamber and say that a particle in this state has property L. Additionally, each one has a randomly assigned position and velocity. Take a macrostate M of the system to be the number of particles to the left of the midline, neglecting which individual ones have property L. A microstate takes note of which particles have property L. In essence, it distinguishes each particle from all the others; imagine the particles are labeled $1, 2, \ldots, 100$ and are marked if they have property L. A macrostate description views the particles as indistinguishable and can be taken to be a number between 0 and 100. Now, we count the number of microstates that correspond to a specified macrostate M. So, to start, the system's macrostate is $M = 100$, which is achieved by one microstate in which every particle has a mark. What happens if exactly one particle moves onto the right side, producing a macrostate of $M = 99$? That particle could be any one of the hundred and there are 100 ways to mark 99 particles, implying that 100 microstates satisfy the given macrostate condition.

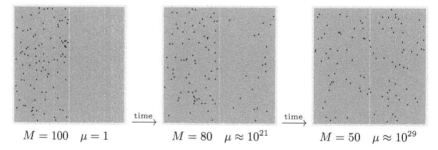

$$M = 100 \quad \mu = 1 \qquad M = 80 \quad \mu \approx 10^{21} \qquad M = 50 \quad \mu \approx 10^{29}$$

Figure 2.13 One-hundred particles in a box. The initial macrostate state (left) is realized by one microstate: all particles have property L. The system evolves to macrostates for which there are successively more microstates.

If the macrostate is $M = 98$, we count the number of microstates with two particles $\{A, B\}$ on the right. For particle A there are 100 possibilities and, for each of those, B can be any of the 99 that remain. The total number of possible pairs is $100 \cdot 99$. However, using particle labels, this value counts both

$\{A = 5, B = 17\}$ and $\{A = 17, B = 5\}$, for example. But, each of these cases corresponds to the same microstate, namely that particles 5 and 17 occupy the right side. What this means is that we've over-counted by a factor of two. Accounting for the excess, there are

$$\frac{100 \cdot 99}{2} = 4950$$

microstates with 98 particles on the left side.

One more instance is enough to see the general form. Given the macrostate $M = 97$ with three particles $\{A, B, C\}$ to the right, A can be any of the 100, for each of which B can be any of the remaining 99, for each of which C can be any of the remaining 98. Taking the product of these three values is an over-count by a factor of $6 = 3 \cdot 2$. As we worked out previously, that's the number of ways three items—A, B, and C in this case—can be shuffled. The resulting microstate is the same no matter how we shuffle the three particles. Therefore, the number of microstates corresponding to macrostate $M = 97$ is

$$\frac{100 \cdot 99 \cdot 98}{3 \cdot 2} = 161,700.$$

In the general case of macrostate M, we can seize a formula for the number of microstates:

$$\mu(M) = \frac{100 \cdot 99 \cdot 98 \cdot \ldots \cdot (M+1)}{(100 - M) \cdot (100 - M - 1) \cdot \ldots \cdot 3 \cdot 2}.$$

Figure 2.14 displays the relationship between μ (the Greek letter μ (mu), often used to abbreviate "micro") and the variable M, which is why we express μ as function of M. The key features here are 1) the steepness of the rise from $M = 40$ to $M = 50$ indicating that the number of microstates grows very fast as the macrostate value increases, 2) the peak occurs at $M = 50$, and 3) the enormous values that occur, especially near the peak. The plot seems to show that μ is barely growing from 0 to 30 and from 70 to 100. The values of μ for $M = 97, 98, 99, 100$ counters this impression, which is an artifact of plotting the huge values of μ near $M = 50$. The curve's reflective symmetry is due to the fact that you can switch the roles of left and right sides for values of M that are equidistant from 50. Take $M = 45$ and $M = 55$, for example. When $M = 55$, we can count the number of ways that $45 = 100 - 55$ particles can be to the right, which is equal to the number of ways 45 particles can be to the left.

The quantity μ is the nub of the concept known as *entropy*. For technical and mathematically aesthetic reasons that aren't important for our purposes, thermodynamical theory takes entropy to be proportional to the logarithm of μ. According to the Second Law of Thermodynamics, given a particle system that's evolving in isolation from energy inputs, the entropy tends to increase. As indicated by the arrows in Figure 2.13, such a tendency can account for the familiar perception of a direction to the passage of time.

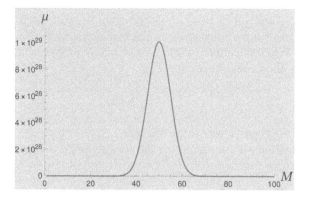

Figure 2.14 Entropy as a graph of microstate number as a function of macrostate.

To see how entropic behavior occurs, consider the case when $M = 98$ for which we find two particles on the right side of the box. Allowing a suitable amount of time to pass for particles on the left to pass to the right and vice-versa, what's the most likely outcome entropy-wise? To keep things simple, assume that the system evolves in such a way that the number of particles to the right changes by at most one. The number of ways the entropy can decrease or remain the same is $5050 = 100 + 4950$. Since there are $161,700$ ways for the entropy to increase, that result is overwhelmingly more likely. In a more realistic setting where the number of particles is on the order of 10^{23}, the likelihood of entropy increase is amplified dramatically.

Following our line of thought indicates that the system will tend to the most probable macrostate, namely $M = 50$, the one for which there is the largest number of microstates. This is not to say that if the system were in macrostate $M = 50$, it couldn't fluctuate out of it. Rather, the second law tells us that finding a system to be far from its maximum entropy state—if it has one— is extremely unlikely. In other words, if the system's entropy is maximal, that quantity will remain close to its maximum value. From this state of affairs, we glean a type of statistical symmetry: when a system of particles near maximum entropy undergoes transformation through interaction, its entropy remains nearly the same.

For a final remark, were we to replace with a crowd of ants a bunch of particles moving according to rules that determine their trajectories, the quest for a fixed property or quantity seems to be a much greater challenge. Unlike the behavior of a mechanical system, the activity of a population of organisms is difficult to forecast even over short spans of time. Something like an ability of the individuals to make choices introduces a degree of complexity not present in agents that are inorganic. Speaking of the power to choose, we're ready to connect that capability to our central problem of solving polynomial equations and symmetry will forge the link.

CHAPTER 3

Geometry of Choice: Symmetry's Cost

3.1 SPACES WHERE THE ROOTS LIVE

We now return to the world inhabited by polynomial equations and their solutions. Recalling the key fact about polynomials that use complex numbers as coefficients, the fundamental theorem of algebra states that such a polynomial can be decomposed into a product of expressions in the first degree:

$$z^n + a_{n-1}z^{n-1} + \cdots + a_1 z + a_0 = (z - z_1)(z - z_2)\ldots(z - z_n).$$

Crucially, the roots z_1, z_2, and so on are also complex numbers.

In an obvious way, the roots reside in the line of complex numbers. Somewhat less obviously, we can take the set of roots to form a collection of points in a higher-dimensional space. For the degree-2 case, form the two-dimensional *space of roots* by taking z_1 and z_2 to be coordinates of points. In other words, each point in the "root space" is given by an ordered pair (z_1, z_2) and thereby, corresponds to the quadratic polynomial

$$(z - z_1)(z - z_2).$$

What is this space that houses pairs of roots? Recall from coordinate geometry that, once we've selected a pair of axes, each location in a plane has a unique address consisting of a pair of real numbers (x, y). Moreover, each real-number pair picks out a unique location. This construction differs from a root space by virtue of the root coordinates being pairs of complex numbers. It makes sense, then, to view the space of roots as a plane of points made manifest by complex-number pairs.

To avoid a possible source of confusion, let's address the matter of the root-space's dimension. On one hand, since it takes two numerical values to specify a location, the dimension can be taken to be two. On the other hand, the coordinates are complex numbers each of which is determined by a pair of real

DOI: 10.1201/9781003098164-3

numbers. Thus, it takes four real numbers to locate a point in the space of roots and so, it seems that the dimension is rightly four. This dimension question is not an actual problem, but comes down to the means of reference that we adopt in describing the space. Mathematicians deal with the situation by distinguishing two types of dimension: real and complex. Figure 3.1 compares the two. The complex line and plane have respective complex dimensions equal to one and two, while their real dimensions—closer to ordinary usage—are two and four.

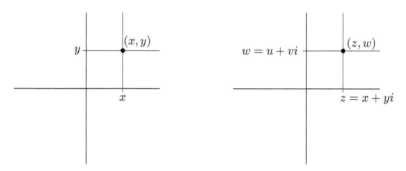

Figure 3.1 Real and complex dimensions. The system of real coordinates on the left can be more or less taken literally. It locates a point in a plane as a pair of real numbers (x, y) each of which lies at the intersection of two real lines. On the right is a system of complex coordinates for which the diagram is a figurative depiction of a space with four real dimensions. Although it can be a challenge to visualize, a point in this space is picked out by intersecting two complex lines—equivalent to real planes—not by crossing real lines as the drawing indicates.

For a third-degree polynomial, the root space is the three-dimensional world (in the complex sense) comprised of ordered triples of complex numbers (z_1, z_2, z_3). Passing to higher degree, ordered collections $(z_1, \ldots, z_n)$ of n complex numbers make up the space of roots for a polynomial of degree n. Mediated by group considerations, the equation-solving algorithm to come will rely on the discovery of structures that realize symmetries exhibited in the root space, an issue to which we now turn.

3.2   SHUFFLING ROOTS AND SOLVING EQUATIONS

What's important to see here, using the factored form, is that the points (z_1, z_2) and (z_2, z_1) correspond to the *same* polynomial. So long as the two roots are unequal, these two points are different. In terms of the symmetry framework, the polynomial is unchanged when the roots undergo a shuffle transformation, namely,

$$\begin{bmatrix} z_1 & z_2 \\ \downarrow & \downarrow \\ z_2 & z_1 \end{bmatrix} \quad \text{which we abbreviate as} \quad \begin{bmatrix} 1 & 2 \\ \downarrow & \downarrow \\ 2 & 1 \end{bmatrix}.$$

After shuffling its roots, a polynomial is the same as the one beforehand. Two polynomials are equal when and only when all of their respective coefficients are identical, implying that a_0 and a_1 are shuffle-invariant. As seen previously, a quadratic naturally gives rise to root equations:

$$z_1 + z_2 = -a_1 \qquad z_1 z_2 = a_0.$$

Consequently, the invariance of the coefficients spreads to the expressions on the left side of these equations—call them *basic root functions*.[1] In this case, it's easy to check that these functions don't vary when the roots are shuffled. The root-exchanging symmetry here can be be depicted graphically, as in Figure 3.2.

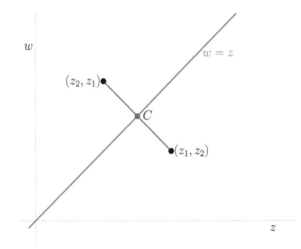

Figure 3.2 Root-space symmetry of a quadratic. The root points (z_1, z_2) and (z_2, z_1) correspond to the same polynomial. Reflecting through the (red) line defined by $w = z$, exchanges these points. The same effect results from rotating by a half-turn about the point C where the mirror line and the (gray) segment between the two root points meet. Taking the reflection along with the trivial transformation produces a group of two elements. Adding the rotation forms a four-element group that, according to Symmetry Principle I, includes the result of combining the half-turn and reflection. The "new" transformation turns out to reflect across the line through both pairs of root points. Using a real number plane gives a cartoon version of what happens in the plane of complex-number pairs.

We can make a nice figurative sketch of what's going on here geometrically. Refer to Figure 3.3 and use (z, w) as coordinates. The red line is the set of points that satisfy the equation $z + w = -a_1$. In blue we have a curve known as a conic section or just conic. Restricting the coordinates to real numbers, the curve is a hyperbola. Its constituent members are solutions to a second

---

[1] In the mathematics literature, they're called elementary symmetric functions the study of which is a centerpiece of the field known as combinatorics.

equation, $zw = a_0$. Let the points in purple (red plus blue) belong to both the line and curve. If one of these is (z_1, z_2), substitution of its coordinates into the defining equations returns the root equations:

$$z_1 + z_2 = -a_1 \qquad z_1 z_2 = a_0.$$

It now follows from the form taken by these equations that the second intersection point also solves both of them and so, must be (z_2, z_1).

When it comes to characterizing the symmetries of a quadratic, the structure here clarifies the matter. Specifically, the transformation that sends a point (z, w) to (w, z) is nothing other than reflection through the black line where $w = z$. The effect of this reflection is to fix the line and conic as collections of points. It also exchanges the two intersections. Under the rotation mentioned in Figure 3.2, the conic curve moves to a different location and so, lacks the character of quadratic symmetry.

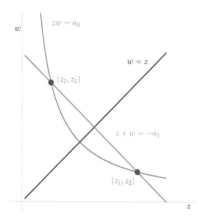

Figure 3.3 Geometry of a quadratic.

Let's see what happens when the degree increases. Having another look at a cubic, its factored expression—hence, its polynomial form—is unchanged after root-shuffling occurs. Since three items are shuffled, the group that acts on the roots z_1, z_2, and z_3 is identical to the six-element set that appears when tracking the vertices of an equilateral triangle as its symmetry group acts on them. Here, again, coefficients a_0, a_1, and a_2 are shuffle-invariant as are the cubic's three basic root functions—the left sides of the root equations derived in Chapter 1:

$$z_1 + z_2 + z_3 = -a_2 \qquad z_1 z_2 + z_1 z_3 + z_2 z_3 = a_1 \qquad z_1 z_2 z_3 = -a_0.$$

As with the quadratic, we can interpret this system of equations in terms of geometry. Each equation determines a surface in three-dimensional complex space. Figure 3.4 conveys the geometric nature of this three-surface configuration. Two of them form a curve when they intersect and the third surface cuts through the curve in a finite set of points. A fundamental theorem pertaining to the geometry of multi-variable polynomials implies that the number of points that reside on all three surfaces is given by the product of the equations' degrees; in this case, they are 1, 2, and 3, respectively. This outcome requires a complex space in which to work, giving a nice illustration of how valuable complex numbers can be. So, we have three surfaces intersecting in six points, as the figure shows. If one of these is (z_1, z_2, z_3), the others are obtained by shuffling transformations:

$$(z_1, z_2, z_3), (z_3, z_1, z_2), (z_2, z_3, z_1), (z_1, z_3, z_2), (z_3, z_2, z_1), (z_2, z_1, z_3).$$

To summarize, associated with a cubic polynomial—its coefficients, to be precise—are three shuffle-invariant surfaces that pick out the polynomial's six root points.

Figure 3.4 We can depict the geometry of a cubic polynomial by considering the intersections of three surfaces associated with its color-matched root equations (left). In the right view, the surfaces have some transparency in order reveal six points that satisfy all three equations and thereby belong to all three surfaces.

Our reasoning extends to polynomials of any degree; shuffling the roots produces no change in the coefficients and associated root functions. In light of this observation that the value taken by a root function is an unchanged property, a group of shuffles qualifies as a symmetry group. What makes the basic root functions for a fixed number of roots worthy of the name 'basic' is the following fact, which will be of service to one of the methods of equation-solving that we consider.

Shuffle Theorem. *Any function of the roots that remains the same when the roots are shuffled in every way can be expressed in terms of the basic root functions.*

A simple example illustrates this effect. The function

$$z_1^2 + z_2^2$$

is unchanged when z_1 and z_2 trade places. So, the theorem tells us that it agrees with some expression using only the basic root functions. To wit:

$$z_1^2 + z_2^2 = (z_1 + z_2)^2 - 2\,z_1 z_2.$$

As things now stand, the connection between a polynomial P and symmetry is evident. Recalling that in order to determine P's roots z_1, z_2, up to z_n, you have to tease them out of its root equations. Such a feat amounts to obtaining a point in the root space that corresponds to P. But, as we've seen, there is not just one point associated with a polynomial. Take the cubic, for instance. In most cases—where the roots are distinct, there are six points in the root space that correspond to a given polynomial. We already met this collection in a geometric setting:

$$(z_1, z_2, z_3), (z_3, z_1, z_2), (z_2, z_3, z_1), (z_1, z_3, z_2), (z_3, z_2, z_1), (z_2, z_1, z_3).$$

From the polynomial's point of view—excusing the anthropomorphic allusion, there's no distinction to be made between these *root points*. Being indistinguishable is a signal of symmetry's influence. Of course, these points result from applying the six shuffles of three items to any one of the triples.

We can extend our discussion of the quadratic and cubic polynomials to higher degree n. The *complete shuffle group* consisting of all shuffles of the n roots serves as a measure of the polynomial's symmetry.[2] As for a geometric interpretation of a degree-n equation, each coefficient corresponds to a *hypersurface*, which you can think of as an $(n-1)$-dimensional set in n dimensions. A pair of hypersurfaces intersect in an object with $n-2$ dimensions, which carves out an $(n-3)$-dimensional space inside a third hypersurface. A cascade of successive hypersurface intersections continues until all that remains is a set of $n!$ root points—the result of completely shuffling $(z_1, \ldots, z_n)$. We'll soon see that the structure exhibited by some of these groups and related ones lead to other types of geometric object that play an instrumental role in solving equations.

That this symmetry group is the key to solving an equation or not was a profound realization made by Évariste Galois around 1830 [34, Ch. 31]. In the modern and highly abstract subject known as Galois Theory, a typical polynomial's symmetry group of shuffles is called a Galois group. Let's dignify this monumental achievement as the **Galois Doctrine**:

> The obstacle to be overcome in solving a root problem is the polynomial's symmetry. Any solution method must, in some way, break this symmetry by distinguishing among the $n!$ root points.

But, how do you go about breaking a polynomial's symmetry? How do you discriminate between things that are inherently indistinguishable? At the most basic level, you make a choice—one that's pure, one for which there is no accounting.

We've arrived at a core understanding: symmetry-breaking is tantamount to choosing. The next two chapters spell out how choices made when solving equations can result from algorithmic methods. Part II then explores the idea that algorithms of these types can be useful in analyzing and, perhaps, in making difficult choices.

---

[2]The common name for such a group is the symmetric group on n objects.

CHAPTER 4

Compute First, Then Choose

In the story of equation-solving, the question was first addressed in algebraic terms. As indicated earlier, the quadratic equation was more-or-less solved in a verbal style. A foundation for modern algebra was laid during the Renaissance period when the introduction of symbolic expressions provided a mechanism for working such a problem. The Renaissance equivalent of a polynomial hardly resembles the form that emerged during the Enlightenment. (See [17].) Nevertheless, solutions to the cubic and quartic (fourth-degree) equations were found by Cardano and Tartaglia in an episode involving cleverness and some intrigue [44, 5]. By the seventeenth-century era of Descartes and Newton, equations conformed to modern style. This chapter develops our first method for obtaining a polynomial's roots. It involves some technical algebra, mostly at a high-school level. We also have occasion to invoke more advanced ideas and theory.

4.1 SIMPLIFYING A POLYNOMIAL

One path toward root-finding was to deploy a potent weapon of mathematical reasoning: transform the general equation into a special form, one for which a solution is at hand. After solving the special equation, reverse the transformation process and arrive at a solution to the general case with which you began. The method found an enthusiastic proponent in E. W. von Tschirnhaus[45], whose program we examine by applying it to polynomials in degrees two and three. This technique of pure algebra stands in stark contrast to the strongly geometric one on which we will ultimately focus in the next chapter. What motivates inclusion of the algebraic approach is the hope that comparison of the two algorithms will illuminate deliberative practices within and beyond mathematics.

A caution here: The strategy that will now unfold involves a somewhat intricate and extensive use of algebraic calculation and the theory of shuffling. Grasping the technical details is not essential for getting the gist of the procedure. However, this level of exposition is the only way to fully describe and appreciate how the procedure works and, more importantly, how it fails.

DOI: 10.1201/9781003098164-4

Take the general polynomial whose degree is equal to n:

$$P(z) = z^n + a_{n-1}z^{n-1} + \cdots + a_1 z + a_0.$$

By introducing a new variable w that's related to z and appealing to the root equations for $P(z)$, Tschirnhaus proposed the creation of another polynomial called a *resolvent* of $P(z)$. His hope is that the new expression can take a special form in which all intermediate terms vanish:

$$R(w) = w^n - t.$$

The parameter t is a function of a_0 through a_{n-1}. Solutions to the equation $R(w) = 0$ are $w = t^{1/n}$ (the nth root of t) for which there are n distinct complex values. This fact is one of the beauties exemplified by the way complex numbers work. In the algorithm's final step, the transformation is reversed in order to express the roots of $P(z)$.

Now, let's work through Tschirnhaus's method applied to a quadratic. For the general degree-two polynomial

$$P(z) = z^2 + a_1 z + a_2,$$

whose roots are z_1 and z_2, recall the root equations:

$$z_1 + z_2 = -a_1 \qquad z_1 z_2 = a_0.$$

The method's operative tool is the *Tschirnhaus transformation* for which we imagine having roots w_1 and w_2 of a new resolvent polynomial $R(w)$ in a new variable w. That is to say,

$$R(w) = (w - w_1)(w - w_2) = w^2 - (w_1 + w_2)w + w_1 w_2.$$

Crucially, the new roots are polynomial functions of the old roots; specifically,

$$w_1 = z_1 + b \qquad w_2 = z_2 + b.$$

The quantity b is a *parameter* that we treat as an unknown until we can determine a value for it that achieves the goal of giving $R(w)$ a simple form that's easily solved. Such an outcome would result if the coefficient of w turns out to be zero so that the new equation to be solved looks like this:

$$R(w) = w^2 - t = 0.$$

Solving this resolvent equation is as easy as taking a square-root:

$$w = \sqrt{t}.$$

But, here we arrive at the crux: as long as it's not zero, any complex number has *two square roots* each of which is the negative of the other. For example,

$$\sqrt{4} = \pm 2 \quad \text{and} \quad \sqrt{-4} = \pm 2i.$$

Accordingly, the resolvent's roots are

$$w_1 = \sqrt{t} \quad \text{and} \quad w_2 = -\sqrt{t}.$$

Eventually, we want t to depend on a_0 and a_1 alone. Looking at the terms in the original expression of $R(w)$, producing the desired effect calls for setting

$$w_1 + w_2 = 0 \quad \text{and} \quad t = -w_1 w_2.$$

By deploying the Tschirnhaus transformation, express $w_1 + w_2$ in terms of z_1, z_2, and b. Make the prescribed substitutions for w_1 and w_2:

$$w_1 + w_2 = (z_1 + b) + (z_2 + b) = 0$$
$$(z_1 + z_2) + 2b = 0.$$

Remembering the root equation $z_1 + z_2 = -a_1$, we can say that

$$-a_1 + 2b = 0.$$

To satisfy this equation, assign the value $\dfrac{a_1}{2}$ to b. In order to work out the value of t and so, solve the resolvent equation, take the following steps:

$$\begin{aligned}
t &= -w_1 w_2 = -(z_1 + b)(z_2 + b) \\
&= -z_1 z_2 - b(z_1 + z_2) - b^2 \\
&= -a_0 - \frac{a_1}{2}(-a_1) - \left(\frac{a_1}{2}\right)^2 \\
&= -a_0 + \frac{a_1^2}{2} - \frac{a_1^2}{4} \\
&= \frac{a_1^2}{4} - a_0 \\
&= \frac{a_1^2 - 4a_0}{4}.
\end{aligned}$$

The solution to the resolvent equation $R(w) = 0$ is now at hand:

$$w_1 = \sqrt{t} = \sqrt{\frac{a_1^2 - 4a_0}{4}} = \frac{\sqrt{a_1^2 - 4a_0}}{2} \quad \text{and} \quad w_2 = -\sqrt{t} = -\frac{\sqrt{a_1^2 - 4a_0}}{2}.$$

In the final step, we obtain the roots of the original polynomial $P(z)$ by reversing the Tschirnhaus transformation to express z_1 and z_2 in terms of w_1 and w_2:

$$z_1 = -b + w_1 \quad \text{and} \quad z_2 = -b + w_2.$$

Complete expressions for the general quadratic's roots appear in the section to follow.

Implementing the Tschirnhaus method in order to solve cubic equations takes a bit of effort. Let z_1, z_2, and z_3 be the roots of the general third-degree polynomial

$$P(z) = z^3 + a_2 z^2 + a_1 z + a_0.$$

As derived in Section 1.3, P's root equations are

$$z_1 + z_2 + z_3 = -a_2 \qquad z_1 z_2 + z_1 z_3 + z_2 z_3 = a_1 \qquad z_1 z_2 z_3 = -a_0.$$

In the key step, define the Tschirnhaus transformation

$$w_1 = z_1^2 + b z_1 + c$$
$$w_2 = z_2^2 + b z_2 + c$$
$$w_3 = z_3^2 + b z_3 + c$$

with parameters b and c to be determined. Next, we imagine that w_1, w_2, and w_3 are the roots of a resolvent:

$$R(w) = (w - w_1)(w - w_2)(w - w_3) = w^3 + k_2 w^2 + k_1 w + k_0.$$

By equating respective coefficients of w^2, w, and w^0, we can see how k_2, k_1, and k_0 depend on w_1, w_2, and w_3:

$$k_2 = -(w_1 + w_2 + w_3)$$
$$k_1 = w_1 w_2 + w_1 w_3 + w_2 w_3$$
$$k_0 = -w_1 w_2 w_3.$$

From here, we glimpse the feature that conceptually underpins Tschirnhaus's approach. Say that we find suitable values for b and c and so obtain solutions w_1, w_2, and w_3 to the resolvent equation $R(w) = 0$. We can then compute values for the roots of the original polynomial $P(z)$ by solving the quadratic Tschirnhaus equations for z_1, z_2, and z_3. The important condition—one that extends to other degrees—is that finding a solution to the target $P(z)$ relies on solving an equation of *lower degree*. Metaphorically, the undertaking is like climbing a ladder in that reaching the next rung (degree) requires having purchase on the current one.

Being functions of w_1, w_2, and w_3 the coefficients k_2, k_1, and k_0 are also expressible in terms of z_1, z_2, and z_3 as well as b and c. Moreover, the values of k_2, k_1, and k_0 don't change when z_1, z_2, and z_3 are shuffled. A calculation that calls on the Shuffle Theorem as well as a piece of advanced math known as invariant theory yields the expression for the coefficient of w^2:

$$
\begin{aligned}
k_2 &= -(w_1 + w_2 + w_3) \\
&= -\left((z_1^2 + b z_1 + c) + (z_2^2 + b z_2 + c) + (z_3^2 + b z_3 + c)\right) \\
&= -(z_1^2 + z_2^2 + z_3^2) - (z_1 + z_2 + z_3)b - 3c \\
&= -(a_2^2 - 2a_1) + a_2 b - 3c.
\end{aligned}
$$

For clarity, an expression in one color in the third line is equal to the expression of the same color in the fourth line. The reader can easily verify the equality between expressions in blue. Making similar maneuvers for the other two coefficients produces comparable albeit more elaborate results:

$$k_1 = a_1 b^2 - 2a_2 bc + 3a_0 b - a_1 a_2 b + 2a_2^2 c - 4a_1 c + a_1^2 - 2a_0 a_2 + 3c^2$$
$$k_0 = -a_1 b^2 c + a_0 b^3 - a_0 a_2 b^2 + a_2 bc^2 - 3a_0 bc + a_1 a_2 bc + a_0 a_1 b - a_2^2 c^2$$
$$+ 2a_1 c^2 - a_1^2 c + 2a_0 a_2 c - a_0^2 - c^3.$$

The precise form of these expressions is not so important. What really matters is that forcing k_2 to be zero requires c to be a function of b in the first-degree:

$$c = \frac{-a_2^2 + 2a_1 + a_2 b}{3}.$$

When we substitute this expression for c into the function equal to k_1, the result is a quadratic polynomial whose variable is b and whose coefficients depend on a_0, a_1, and a_2:

$$k_1 = \frac{1}{3}(3a_1 - a_2^2)b^2 + \frac{1}{3}(2a_2^3 - 7a_1 a_2 + 9a_0)b + \frac{1}{3}(-a_2^4 + 4a_1 a_2^2 - 6a_0 a_2 - a_1^2).$$

This step is crucial in order for the Tschirnhaus method to be effective. The equation $k_1 = 0$ has lower degree than the cubic whose roots we seek, and as such, its solution is something that we know how to find.

Satisfying the equations $k_2 = 0$ and $k_1 = 0$ determines specific values—call them B and C—for b and c:

$$B = \frac{-2a_2^3 + 7a_1 a_2 - 9a_0 \pm \sqrt{-3D_3}}{2(3a_1 - a_2^2)}$$

$$C = \frac{12a_1^2 - 3a_1 a_2^2 - 9a_0 a_2 + a_2 \pm \sqrt{-3D_3}}{6(3a_1 - a_2^2)}$$

$$D_3 = -4a_1^3 + a_1^2 a_2^2 + 18a_0 a_1 a_2 - 4a_0 a_2^3 - 27a_0^2.$$

Here, the symbol D_3 stands for the *discriminant* of the cubic—hence the subscript 3. We'll soon see how this quantity plays an important role in the symmetry-breaking that's realized by the cubic-solving procedure. In addition, when B and C are substituted for b and c, the form of k_0 becomes one that depends only on the coefficients a_2, a_1, and a_0. Let's denote the resulting value K_0. It turns out that

$$K_0 = \frac{-2D_3\sqrt{-D_3}}{81\sqrt{3}a_0 + 3\sqrt{3}a_2(9a_1 - 2a_2^2) + 27\sqrt{-D_3}}.$$

By arranging for k_2 and k_1 to vanish, Tschirnhaus's goal is achieved. With k_0 expressed as K_0 as well, the cubic resolvent has the special form

$$R(w) = w^3 + K_0,$$

the roots of which are the three cube-roots of $-K_0$. Using the complex numbers

$$V_1 = -\frac{1}{2} + \frac{\sqrt{3}}{2}i \qquad V_2 = -\frac{1}{2} - \frac{\sqrt{3}}{2}i$$

and letting $w_1 = \sqrt[3]{-K_0} = -\sqrt[3]{K_0}$ be any one of the cube-roots, the other two roots of $R(w)$ are

$$w_2 = -V_1 \sqrt[3]{K_0} \quad \text{and} \quad w_3 = -V_2 \sqrt[3]{K_0}.$$

Inserting these factors works because $V_1^3 = V_2^3 = 1$. In the complex numbers, there are three distinct cube-roots of 1 or any number other than 0.

The final step leads to the roots z_1, z_2, and z_3 of the original polynomial $P(z)$. Their respective formulas are given below.

4.2 SOLUTIONS FROM A FORMULA AND A CHOICE

Substituting the derived values for b, w_1 and w_2 reveals the roots of a quadratic:

$$z_1 = -\frac{a_1}{2} + \frac{\sqrt{a_1^2 - 4a_0}}{2}$$

$$z_2 = -\frac{a_1}{2} - \frac{\sqrt{a_1^2 - 4a_0}}{2}.$$

In song, verse, and chant, students in algebra classes the world over have a variant of this *quadratic formula* drilled into their memories. The significance of the red plus and minus signs will be explained presently.

Moving on to the cubic, consider the first Tschirnhaus equation:

$$w_1 = z_1^2 + bz_1 + c.$$

To obtain z_1 in terms of $P(z)$'s coefficients, replace w_1, b, and c by the expressions $-\sqrt[3]{K_0}$, B, and C just derived and then solve the resulting quadratic for z_1:

$$z_1^2 + Bz_1 + C + \sqrt[3]{K_0} = 0.$$

Deploy the formula to arrive at

$$z_1 = \frac{-B}{2} + \frac{\sqrt{B^2 - 4(C + \sqrt[3]{K_0})}}{2}.$$

Substitute the explicit expressions in terms of a_2, a_1, and a_0. After setting

$$G = -2a_2^3 + 9a_1a_2 - 27a_0 \qquad \text{and} \qquad H = a_2^2 - 3a_1$$

we get

$$D_3 = \frac{4H^3 - G^2}{27}.$$

Making further algebraic maneuvers, the roots take the form[1]

$$z_1 = -\frac{a_2}{3} + \frac{H \cdot 1}{3} \sqrt[3]{\frac{2}{G \pm \sqrt{-27 D_3}}} + \frac{1}{3 \cdot 1} \sqrt[3]{\frac{G \pm \sqrt{-27 D_3}}{2}}$$

$$z_2 = -\frac{a_2}{3} + \frac{H \cdot V_1}{3} \sqrt[3]{\frac{2}{G \pm \sqrt{-27 D_3}}} + \frac{1}{3 \cdot V_1} \sqrt[3]{\frac{G \pm \sqrt{-27 D_3}}{2}}$$

$$z_3 = -\frac{a_2}{3} + \frac{H \cdot V_2}{3} \sqrt[3]{\frac{2}{G \pm \sqrt{-27 D_3}}} + \frac{1}{3 \cdot V_2} \sqrt[3]{\frac{G \pm \sqrt{-27 D_3}}{2}}.$$

Curiously, algebra classes aren't expected to memorize the cubic formula.

For our purposes, the key feature of the quadratic and cubic formulas are the red bits. From a purely mathematical point of view, the two roots of a quadratic are indistinguishable, an arrangement that exemplifies the polynomial's two-fold symmetry. Recall that in order to actually solve a quadratic equation, you must break its symmetry. The formula makes it clear that you can accomplish this feat by *choosing* one of the signs—either plus or minus. Such a binary choice adapts the situation faced by Buridan's mule with the two formulas for the roots replacing the identical hay bales. Solving either problem requires selecting one of the options when there's no good reason—indeed, no reason whatsoever—for doing so.

Unsurprisingly, the degree-three equation has more going on than does the second-degree case. Obtaining a root of the cubic again calls for a selection to be made. Specifically, you must choose among the three options 1, V_1, and V_2 each of which corresponds to one solution. However, a further binary choice, indicated by $\pm$, arises from the square-root. You could have elected to take the negative square-root instead. As a consequence, for each of the three picks 1, V_1, and V_2, there are two more—either $+$ and $-$. So, it looks like we have a total of six pairs of choices:

$$(1, +), \ (1, -), \ (V_1, +), \ (V_1, -), \ (V_2, +), \ (V_2, -).$$

(Color coding has been relaxed.) How can this be when there are but three roots? Resolution of this quandary can be found by observing that the six cases form three pairs, each of which give the same root as follows:

$$\underbrace{(1, +), (V_2, -)}_{z_1} \quad \underbrace{(V_1, +), (V_1, -)}_{z_2} \quad \underbrace{(V_2, +), (1, -)}_{z_3}.$$

Lurking here is a deeper issue. When discussing root-shuffling, we saw that a cubic polynomial has six symmetries, namely, all of the possible ways that

---

[1]This is how the computer algebra system *Mathematica* represents the roots.

three roots can be shuffled. By virtue of the cubic's symmetry group, each root is indistinguishable from the other two. In the cubic-formula context, the same goes for the three combinations of two choice-pairs.

For amusement, we can adapt the scenario of Buridan's mule to the cubic equation. As pictured in Figure 4.1, the mule can follow one of six routes (choice-pairs) to reach one of three hay bales (roots). Each route involves making two symmetry-breaking choices (analogous to the calculation of a cube-root and a square-root).

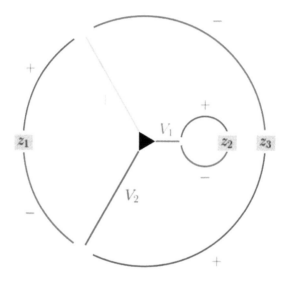

Figure 4.1 Buridan's mule solves a cubic. Starting at the central triangle, the mule first selects one of the three paths marked 1, V_1, and V_2. The chosen course then leads to a split into two trails marked plus and minus. The mule opts to take one of these and thereby arrives at a bale. Each route corresponds to one of the pairs of choices held out by the cubic formula. Observe that the mule can reach each bale by taking two distinct routes.

With the Tschirnhaus method in hand, we could have a go at solving the quartic (fourth-degree) equation. The result would be a formula like those obtained in the quadratic and cubic cases. In this instance, expressing the roots involves the computation of square-roots and cube-roots. Since doing so would call for a sizable proliferation in technical considerations and derived formulas without a reward of additional insight, we refrain from the endeavor.

Polynomials of degrees two, three, and four have solutions the expression of which requires only basic arithmetic operations and root-taking—known as "solution by radicals." A natural query concerns whether the pattern continues into the fifth-degree and beyond. In a momentous twist, our line of thought leads to a negative outcome. Before broaching the topic, we require an important piece of theory related to groups that are companions of polynomials.

4.3   REDUCING A POLYNOMIAL'S SYMMETRY

We return now to a topic noted previously, one which plays an important role in the upcoming chapter. Recall that the quadratic formula includes an expression housed within a square-root—referred to as the quadratic discriminant:

$$D_2 = a_1^2 - 4a_0.$$

Using the root functions

$$a_1 = -(z_1 + z_2) \quad \text{and} \quad a_0 = z_1 z_2$$

to replace the coefficients with expressions in terms of the roots gives an interesting result

$$D_2 = (z_1 - z_2)^2.$$

What makes the discriminant worth mentioning is tied up with its symmetry properties. The shuffle group of the quadratic either leaves the roots alone or exchanges them. Applying either of these actions has no effect on D_2. Notice that the Shuffle Theorem turns this process around: since D_2 is unchanged under all shuffles of two roots, it can be written in terms of root functions.

The same cannot be said of the discriminant's square-root for which we get two values:

$$\sqrt{D_2} = \pm(z_1 - z_2).$$

Whichever quantity we choose—either plus or minus, its sign changes when the roots switch places. In more formal terms, let S_1 be the trivial shuffle that fixes each root and S_2 the one that exchanges them. Then

$$\sqrt{D_2} \xrightarrow{S_1} \sqrt{D_2} \qquad \sqrt{D_2} \xrightarrow{S_2} -\sqrt{D_2}.$$

The important point here is that $\sqrt{D_2}$ remains unchanged under application of half of the quadratic's symmetry group.

Similar behavior occurs when considering the cubic. After transforming the general equation to a special one—the Tschirnhaus resolvent, an expression called the cubic discriminant appears within a square-root:

$$D_3 = -4a_1^3 + a_1^2 a_2^2 + 18a_0 a_1 a_2 - 4a_0 a_2^3 - 27a_0^2.$$

Replace the coefficients with their values found in the root functions

$$a_2 = -(z_1 + z_2 + z_3) \qquad a_1 = z_1 z_2 + z_1 z_3 + z_2 z_3 \qquad a_0 = -z_1 z_2 z_3.$$

The result is
$$D_3 = (z_1 - z_2)^2 (z_1 - z_3)^2 (z_2 - z_3)^2.$$

Here, again, apply any of the six shuffles of the three roots to D_3 and the function remains the same.

Passing to the square-root of D_3, the function
$$\sqrt{D_3} = \pm(z_1 - z_2)(z_1 - z_3)(z_2 - z_3)$$

is unchanged when a certain half of the complete shuffle group is applied and changes sign under the other half. Recall our enumeration in Table 2.1 of the shuffles associated with the cubic.

The action of those six shuffles on the square-root of D_3 is as follows[2]:

$$\sqrt{D_3} \xrightarrow{I} \sqrt{D_3} \qquad \sqrt{D_3} \xrightarrow{R} \sqrt{D_3} \qquad \sqrt{D_3} \xrightarrow{R^2} \sqrt{D_3}$$
$$\sqrt{D_3} \xrightarrow{A} -\sqrt{D_3} \qquad \sqrt{D_3} \xrightarrow{B} -\sqrt{D_3} \qquad \sqrt{D_3} \xrightarrow{C} -\sqrt{D_3}.$$

For instance, compare the behavior of $\sqrt{D_3}$ when R and C act on it:

$$\sqrt{D_3} = (z_1 - z_2)(z_1 - z_3)(z_2 - z_3) \xrightarrow{R} (z_2 - z_3)(z_2 - z_1)(z_3 - z_1)$$
$$= (z_2 - z_3)[-(z_1 - z_2)][-(z_1 - z_3)]$$
$$= (z_1 - z_2)(z_1 - z_3)(z_2 - z_3) = \sqrt{D_3}$$

$$\sqrt{D_3} = (z_1 - z_2)(z_1 - z_3)(z_2 - z_3) \xrightarrow{C} (z_3 - z_2)(z_3 - z_1)(z_2 - z_1)$$
$$= [-(z_2 - z_3)][-(z_1 - z_3)][-(z_1 - z_2)]$$
$$= -(z_1 - z_2)(z_1 - z_3)(z_2 - z_3) = -\sqrt{D_3}.$$

Notice the number of minus signs (in red) that appear when a shuffle is applied: two for R and three for C. It's easy to check that half of the six shuffles— namely, I, R, and R^2—produce an even number of minus signs and the other half—A, B, and C—yield an odd number.

This halving phenomenon occurs in the context of a complete shuffle group for any number of roots $z_1, \ldots, z_n$. The discriminant of the general degree-n polynomial is the product of the squares of the differences between the roots:

$$D_n = (z_1 - z_2)^2 (z_1 - z_3)^2 \cdots (z_{n-1} - z_n)^2.$$

As in the quadratic and cubic cases, D_n is unchanged when any rearrangement of the roots occurs and so, the Shuffle Theorem implies that D_n is expressible in terms of the root functions which, of course, can be formulated using the coefficients $a_{n-1}, \ldots, a_0$.

---

[2]Take note that here we're using the symbols B and C to represent shuffles, not the special values of the cubic coefficients b and c.

The collection of shuffles that generate an even number of minus signs when applied to

$$\sqrt{D_n} = \pm(z_1 - z_2)(z_1 - z_3)\cdots(z_{n-1} - z_n)$$

makes no change in this function. Hence, Symmetry Principle II requires it to be a group. Let's call it the *even-shuffle group*[3] A_n. Although deeper theory in the subject known as combinatorics can be brought to bear on shuffle groups, the vital take-away for our pursuit is that A_n accounts for half the members in a full shuffle group that we denote S_n. The other half consists of *odd* shuffles that make an odd number of minus signs. Note that, when two odd shuffles are applied in succession, they form an even shuffle—a fact not unrelated to the Two Reflections Principle.

Our investigation into equation-solving that relies on geometric structure will call on the even-shuffle groups A_4, A_5, and A_6. By way of illustration, recalling that the group S_4 contains 24 shuffles, the twelve elements of A_4 appear in Table 4.1. Verifying that each of these shuffles gives rise to an even number of minus signs in $\sqrt{D_4}$ is a simple albeit tedious matter.

Table 4.1 Twelve Even Shuffles in S_4 Form the Group A_4.

$$
\begin{bmatrix}1&2&3&4\\ \downarrow&\downarrow&\downarrow&\downarrow\\ 1&2&3&4\end{bmatrix}
\begin{bmatrix}1&2&3&4\\ \downarrow&\downarrow&\downarrow&\downarrow\\ 2&1&4&3\end{bmatrix}
\begin{bmatrix}1&2&3&4\\ \downarrow&\downarrow&\downarrow&\downarrow\\ 3&4&1&2\end{bmatrix}
\begin{bmatrix}1&2&3&4\\ \downarrow&\downarrow&\downarrow&\downarrow\\ 4&3&2&1\end{bmatrix}
$$

$$
\begin{bmatrix}1&2&3&4\\ \downarrow&\downarrow&\downarrow&\downarrow\\ 2&3&1&4\end{bmatrix}
\begin{bmatrix}1&2&3&4\\ \downarrow&\downarrow&\downarrow&\downarrow\\ 3&1&2&4\end{bmatrix}
\begin{bmatrix}1&2&3&4\\ \downarrow&\downarrow&\downarrow&\downarrow\\ 2&4&3&1\end{bmatrix}
\begin{bmatrix}1&2&3&4\\ \downarrow&\downarrow&\downarrow&\downarrow\\ 4&1&3&2\end{bmatrix}
$$

$$
\begin{bmatrix}1&2&3&4\\ \downarrow&\downarrow&\downarrow&\downarrow\\ 3&2&4&1\end{bmatrix}
\begin{bmatrix}1&2&3&4\\ \downarrow&\downarrow&\downarrow&\downarrow\\ 4&2&1&3\end{bmatrix}
\begin{bmatrix}1&2&3&4\\ \downarrow&\downarrow&\downarrow&\downarrow\\ 1&3&4&2\end{bmatrix}
\begin{bmatrix}1&2&3&4\\ \downarrow&\downarrow&\downarrow&\downarrow\\ 1&4&2&3\end{bmatrix}
$$

In order to discern the significance of the discriminant and its square-root, we revisit the quadratic and cubic formulas. We found that the roots of a degree-two polynomial are captured by

$$\frac{-a_1 \pm \sqrt{D_2}}{2}.$$

Extracting one of the square-roots of D_2 breaks the two-fold quadratic symmetry. Having completely reduced the polynomial's symmetry, a root is at hand. So states the Galois Doctrine.

As for the third-degree polynomial, computation of a square-root of D_3 reduces the symmetry to be overcome from the six-fold full shuffle group S_3 to that of three-fold group A_3. The reduction in symmetry is evident from the

---

[3]In standard nomenclature, the even permutations make up the *alternating* group.

cubic formula according to which a root can be expressed in the form

$$-\frac{a_2}{3} + \frac{H \cdot X}{3} \sqrt[3]{\frac{2}{G + \sqrt{-27D_3}}} + \frac{1}{3 \cdot X} \sqrt[3]{\frac{G + \sqrt{-27D_3}}{2}}$$

with X equal to 1, V_1 or V_2. By computing $\sqrt{D_3}$, you break the two-fold symmetry embodied by the formula. Only the cube-root remains to be found, an effort that requires three-fold symmetry to be broken.

In the general case of a degree-n equation with symmetry group S_n, Galois Theory works out a process by which the symmetry that a solution-procedure must overcome is halved. Being in possession of the discriminant's square-root furnishes a tool with which you can construct a Tschirnhaus resolvent whose symmetry group is A_n. In effect, we're enlarging an equation's set of available numbers, which absorbs some of its associated symmetry. As in the cases examined previously, you can derive a root of the original polynomial from a solution to the resolvent equation. Being able to work with the even-shuffle groups will be of service when it comes to the culmination of the book's first part: the development of an equation-solving method that's rooted in geometry and dynamics. Before setting out on that part of the journey, we have one more piece of the pure algebra puzzle to consider; namely, when the degree is five, how do we know that we can solve the resolvent equation? In short, we don't.

4.4 WHAT GOES WRONG

Turning to the watershed case of the fifth-degree, write the quintic polynomial as we did the quadratic and cubic—that is, factored into five pieces each of which calls on one of the roots z_1 to z_5:

$$p(z) = z^5 + a_4 z^4 + a_3 z^3 + a_2 z^2 + a_1 z + a_0$$
$$= (z - z_1)(z - z_2)(z - z_3)(z - z_4)(z - z_5).$$

Here, again, the coefficients a_4, a_3, a_2, a_1, and a_0 present us with the root functions that remain the same when the roots are shuffled:

$$a_4 = -(z_1 + z_2 + z_3 + z_4 + z_5)$$
$$a_3 = z_1 z_2 + \cdots + z_4 z_5$$
$$a_2 = -(z_1 z_2 z_3 + \cdots + z_3 z_4 z_5)$$
$$a_1 = z_1 z_2 z_3 z_4 + \cdots + z_2 z_3 z_4 z_5$$
$$a_0 = -z_1 z_2 z_3 z_4 z_5.$$

Although the quintic's symmetries consist of the full shuffle group S_5, we take a significant step by reducing the symmetry group to just the even shuffles A_5. As witnessed when simplifying the quadratic and cubic according to the

Tschirnhaus procedure, this reduction creates a fifth-degree resolvent after computing the square-root of the quintic's discriminant D_5.

Why is it important to work with A_5? It turns out that the even-shuffle group has a special property that blocks the simplification envisioned by Tschirnhaus. Furthermore, we'll see in the next chapter that the geometry associated with A_5 is exceptional and provides a nice way to describe this property.

Without exhibiting the technical details in their formidable entirety, let's outline how to carry out a Tschirnhaus process for solving the quintic. Begin with the transformation

$$w_1 = z_1^4 + bz_1^3 + cz_1^2 + dz_1 + e$$
$$w_2 = z_2^4 + bz_2^3 + cz_2^2 + dz_2 + e$$
$$w_3 = z_3^4 + bz_3^3 + cz_3^2 + dz_3 + e$$
$$w_4 = z_4^4 + bz_4^3 + cz_4^2 + dz_4 + e$$
$$w_5 = z_5^4 + bz_5^3 + cz_5^2 + dz_5 + e,$$

from which a resolvent forms:

$$R(w) = (w - w_1)(w - w_2)(w - w_3)(w - w_4)(w - w_5)$$
$$= w^5 + k_4 w^4 + k_3 w^3 + k_2 w^2 + k_1 w + k_0.$$

Each coefficient k_4 to k_0 is defined in terms of w_1 to w_5 and then, after applying the Tschirnhaus transformation, in terms of the roots z_1 to z_5 as well as the parameters b, c, d, and e. Because k_4 to k_0 are unchanged when z_1 to z_5 shuffle, they can be rewritten as expressions in a_0 to a_4. Adopting Tschirnhaus's strategy means that we impose the conditions

$$k_4 = 0, \quad k_3 = 0, \quad k_2 = 0, \quad k_1 = 0.$$

In the event that we succeed in satisfying them, the resolvent equation takes an easily-solved form:

$$w^5 + k_0 = 0.$$

The five solutions w_1 to w_5 of this equation are complex numbers—the fifth-roots of $-k_0$. From there, we can substitute these results into the Tschirnhaus equations and solve the five resulting quartics for z_1 to z_5. What could go wrong?

In order to make k_4 to k_1 disappear, we must determine specific values for b, c, d, and e. That is, each parameter is a function that depends on a_0 to a_4. As with the cubic, we can take the equations $k_4 = 0$ to $k_1 = 0$ in succession, eliminating one parameter at each step.

Begin with k_4. The parameter b appears to the first power and no higher. Hence, you can cause k_4 to vanish by solving for b in $k_4 = 0$. Afterward, b becomes an expression in c, d, and e as well as a_0 to a_4. Next, substitute the value just derived for b into k_3 to obtain a form in which the highest power of c is two. Use the quadratic formula to determine a value for c in terms of d and e that satisfies $k_3 = 0$. Of course, the value for c that we obtain involves a square-root. After substituting this value for c into the coefficient k_2, you then confront the equation $k_2 = 0$ in which the square-root term resides.

Here you find that Tschirnhaus's ladder is missing a couple rungs. Obtaining values for d or e that make $k_2 = 0$ amounts to solving a *sixth-degree* equation in either variable. You might wonder how this can be, since k_2 contains only powers of d and e that are less than four. Evidently, this fact underpins the goal of climbing the Tschirnhausian ladder, in that a solution to the quintic can be achieved by solving only equations whose degree is less than five. But, the plan fails in historic poignancy: for the quintic there's no general formula involving standard algebraic operations—including square-roots, cube-roots, and fifth-roots.

Using a toy example, let's illustrate how this phenomenon can occur. Take A and B to be variables and consider the equation

$$A^3 + 2B^3 - 3\sqrt{A^2 + B^3} = 0,$$

noting that no term has a power greater than three. Reassemble the items as follows

$$A^3 + 2B^3 = 3\sqrt{A^2 + B^3}$$

and then square each side:

$$A^6 + 4A^3B^3 + 4B^6 = (A^3 + 2B^3)^2 = 9(A^2 + B^3).$$

We end up with an equation whose degree is six no matter which variable is used:

$$A^6 + 4B^3A^3 - 9A^2 + 4B^6 - 9B^3 = 0.$$

Tschirnhaus's approach to solving equations is an exercise in algebraic manipulation. Ultimately, its failure when applied to the fifth-degree case stems from of the special nature possessed by the symmetry group A_5. Galois discovered this key to the quintic and beyond through the development of abstract group theory. Maybe it seems that with this outcome equation-solving comes to a halt. However, why should a symbolic formula be the only route to this end? Next, we lay out the book's centerpiece: a general procedure that produces a polynomial's roots. When worked out for quintics, the crucial resource on which this method draws is the special *geometric* nature of A_5.

CHAPTER 5

Choose First, Then Compute

> If someone says ... that mathematical reasoning cannot be carried beyond a certain point, you may be sure that the really interesting problem begins precisely there.

> –Felix Klein [33]

Following Galois's discoveries (*circa* 1830), the study of polynomial equations fell into a somewhat dormant state. About fifty years later, Felix Klein brought a novel perspective to bear on the subject. In a tour de force of articles (for instance, [29]), he proposed a program for conducting research into the problem of solving equations. His classic monograph, *Lectures on the Icosahedron*, developed a means by which the roots of a quintic can be determined [32].

At the foundation of an equation-solving procedure, Klein joined a polynomial's symmetry group to a suitable geometric space. The offspring of this marriage between algebra and geometry guides the steps that we now begin to take [32, p. 138].

Klein's Rule. *When attempting to solve a polynomial equation $P = 0$, look for a space of the smallest possible dimension where P's symmetry group can be realized as a collection of transformations.*

Our first task is to explain what this exhortation means and how it can be accomplished. Before doing so, this is an apt place to remark on what motivates the current approach to equation-solving. Klein's Rule is not essential to the effort as there are methods of solution that fail to adhere to it. Such procedures tend to focus on speed and efficiency. In contrast, we appreciate the precept as an aesthetic criterion for creating root-finding algorithms clothed in elegant geometry.

5.1 A LINE THAT BECOMES A SPHERE

As seen earlier, the line of complex numbers provides a natural residence for a polynomial's coefficients as well as its roots. We noted that the complex line takes the form of a plane in the standard or real sense of the term, which we use for the moment. Applying a cartographic technique called stereographic projection, this plane surface wraps around a sphere in a way that matches each point in the plane with exactly one on the sphere—with a single exception. Worthy of note is the fact that you cannot accomplish this type of wrapping with a sheet of paper; it requires an elastic material that can shrink and stretch, as you see when inflating a ball.

Figure 5.1 illustrates how this projection works. Submerge a sphere halfway into a plane so that the northern hemisphere is above the plane, the southern hemisphere is below, and the equatorial circle is where the plane and sphere intersect. To be specific, take the sphere's radius to be one unit. Imagine a laser at the north pole N aimed at a point S on a transparent sphere. Ignoring refractive effects, the beam hits the plane at a point L. Extending this correspondence over the entire plane creates a pairing between every point on the plane and all but one point on the sphere. The lone omission is the north pole itself. A remedy for this mismatch is to add a single point at infinity to the plane and pair it with N. Note that the projection process—for obvious reasons, called a mapping or just *map*—allows passage from the sphere to the plane as well as the reverse, from the plane to the sphere.

To illustrate what the stereographic map does, take the sphere's center to be the complex number 0 which is paired with the south pole. Each point on the equatorial circle pairs with itself while points in the northern/southern hemisphere are mapped to the exterior/interior of the equatorial circle.

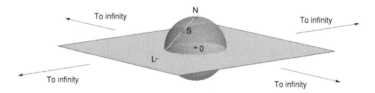

Figure 5.1 Stereographic projection between the complex line (real plane) and a sphere. A green laser beam shines from the north pole N through a point S on the sphere and reaches the complex line at a point L. We say that S maps to L and vice versa, meaning that the projection process is reversible. Note that any radial path away from 0 leads to a lone point at infinity that pairs with N by virtue of horizontal lines through the north pole "pointing to" the same infinitely removed location.

The dividend paid by investing in stereographic projection is that we can look for relevant structures on the *complex sphere* in addition to the complex line. Moreover, when convenient, we can pass back and forth between the two models of space.

5.2 SYMMETRICAL STRUCTURES

Our next challenge is to comply with Klein's Rule by detecting spaces of least dimension where structural equivalents to polynomial symmetry groups appear. The journey begins with the group of a quadratic polynomial and continues by incrementing the degree. At each step we identify and examine a geometric structure that manifests the symmetries associated with the polynomial in question. Recall that for an equation whose degree is n, the relevant groups are S_n or A_n consisting respectively of all shuffles or the half that are even. To ensure that the dimension is as small as possible, first consider whether the respective group of symmetries presents itself in one-dimensional space, either in the complex line or sphere. If not, take a look in two dimensions, and so on.

As a general strategy, we seek a group of transformations that move n objects in precisely the same way that S_n or A_n shuffles n roots. Because of the abstract quality of groups, what counts as an "object" can vary from case to case. What matters is that the two groups agree structurally. For small values of n, the structure we seek appears as a *polyhedron*: a two-dimensional figure residing in three real dimensions that's analogous to a polygon in two dimensions. Specifically, you build a polyhedron by attaching polygons (faces) along their edges in such a way that the resulting form closes up with no gaps between faces. Familiar examples are a cube, prism, and Egyptian pyramid. Imagining that the faces can shrink and stretch but can't be cut, many polyhedra—including the examples just noted—can be deformed into a spherical shape. Such "spherical polyhedra" can be viewed as networks of edges on the complex sphere.

The geometry associated with a polynomial's group provides the crucial link between the impediment to computing a root—the Galois Doctrine—and the device that we use to clear that hurdle. It might be said that we're using a geometric version of Galois Theory to approach the problem of solving equations.

$n = 2$

As noted in Section 3.2, a degree-two polynomial has a group of symmetries S_2 that contains just two elements one of which is trivial. Of course, the relevant behavior is exhibited in a space whose complex dimension is two (coordinates occur as pairs of complex numbers) and whose real dimension is four (a quadruple of real numbers specifies a location). Here, getting to a lower dimension is particularly easy. We can satisfy Klein's Rule by simply noticing that the behavior of S_2 can play out in one dimension—that is, on the complex line. Figure 5.2 treats both situations while revealing another mathematical interpretation of the choice before Buridan's mule.

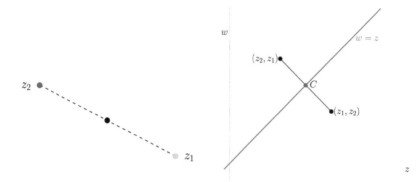

Figure 5.2 S_2 symmetry of a quadratic. The left-hand diagram indicates a rotational motion on the complex line that exchanges the roots. For comparison, reproduced on the right is an illustration of geometric shuffling through reflection applied to pairs of a quadratic's roots. There's an apparently similar two-fold rotational symmetry that also can be said to swap z_1 and z_2. However, the environments in which the respective events take place are different. On the left, the roots reside in a *literal* space where an actual rotation moves each one to the location of the other. The behavior depicted on the right side occurs in an *allegorical* space of two real dimensions—one that merely suggests a four-dimensional setting. The transformation that exchanges a point's complex coordinates—turning (z, w) into (w, z)—is a reflection in the complex sense.

$n = 3$

In Section 2.4 we saw that the complete shuffle group S_3 for three roots is structurally equivalent to an equilateral triangle's set of symmetries. This observation suggests a way to construct a polyhedron whose symmetry group is also equivalent to S_3. Think of the triangle as the base of two pyramids, as pictured in Figure 5.3. Inscribing the double pyramid in a sphere and then, from the sphere's center, projecting it onto the spherical surface produces a beach-ball structure with the same group of rotational symmetries as the polyhedron. Figure 5.4 indicates how this projection looks.

As for the beach-ball's rotational symmetry, the spherical image on the left side of Figure 5.5 shows four axes. Between the north and south poles a three-fold axis appears while there are three two-fold axes passing through the distinguished equatorial points—give them color symbols R, B, and G. In fact, we can take those points to be the three objects that the symmetries shuffle. Revisiting Table 2.1, the collection of symmetries here is familiar. If you substitute 1, 2, and 3 for R, B, and G, you get the complete shuffle group S_3 and so, obey Klein's Rule.

On the figure's right is the configuration of vertices and edges as it appears on the complex line after stereographic projection. Alternatively, the part within the circle is what the beach ball looks like if viewed from above the north pole. When the sphere undergoes a rotational symmetry, points on the

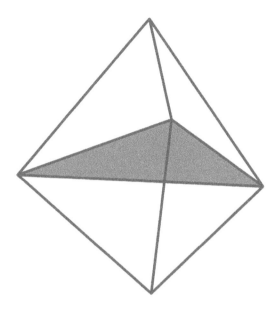

Figure 5.3 Double triangular pyramid.

line move accordingly. In each setting, the collection of edges remains the same—as an aggregate, not as individuals. However, the transformation on the line can be, but need not be, a rotation in the standard sense of turning about a central point. In each space, turning about the orange axis by one-third of a complete revolution counts as a symmetry. Applying a half-turn about one of the equatorial axes corresponds to a transformation of the complex line that's not a standard rotation on a plane. Nor is it a reflection. For evidence, rotation about the red axis on the sphere corresponds to a transformation on the line that fixes the red vertex, exchanges the green and blue vertices, and, somewhat dramatically, sends the points inside/outside the circle to the points outside/inside the circle. The function on the line that a rotation of the sphere engenders belongs to the class of *Mobius transformations*—a collection of functions from which rich and beautiful geometry emerges [38].

$n = 4$

So far, we've identified spaces in which individual points in a collection act as objects that are completely shuffled when acted upon by the collection's rotational symmetry group. If you look for four points—denoted A, B, C, and D—on the sphere that experience complete shuffling when a group of rotations is applied to them, you run into difficulties. Suppose you can rotate

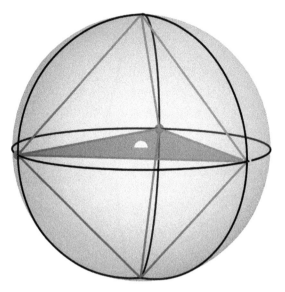

Figure 5.4 Double triangular pyramid as a network on a sphere. Lying in the plane of the shaded triangle, the equatorial circle is divided into thirds by three colored points. With a point-source of light at the sphere's center, the gray edges of the double pyramid cast shadows (in black) on the spherical surface.

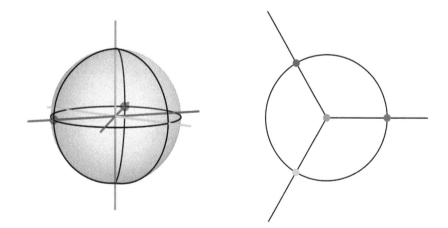

Figure 5.5 Two views of the double triangular pyramid that reveal S_3 as its symmetry group.

the sphere so that the points are "cycled" as follows:

$$A \to B \to C \to D$$

The notation here is a simplified version of the way we expressed shuffles previously. It must be that they all lie in a plane that's perpendicular to the turning axis. But then, no rotation can produce the cycle

$$A \to B \to D \to C$$

which is one of the shuffles in S_4. Similar considerations exclude even shuffles as well. A set of four *points* is too simple for the purpose.

As an alternative to individual points being shuffled objects, what about taking more than one point to be an object? A cube has eight vertices that come in four opposite pairs known as *antipodes*. Figure 5.6 shows a spherical cube with the pairing of antipodal vertices indicated by diagonal line segments in four colors that we abbreviate as R, B, G, Y. Also pictured is an octahedron, which can be thought of as two pyramids that share a square base. These two polyhedra are closely associated. Place a vertex in the center of each face of the cube and, if the faces are adjacent, connect the vertices with an edge. This process makes a new polyhedron which happens to be the octahedron. Because the construction of the octahedron respects all of the cube's rotational and reflective symmetries, the two figures enjoy the same symmetry group.

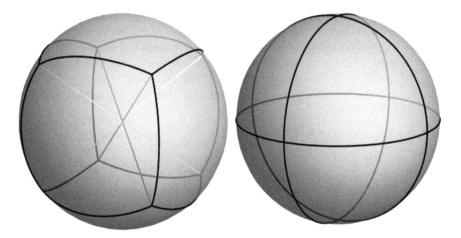

Figure 5.6 Two polyhedra each of whose rotational symmetry group is S_4. At left is a cube with its four diagonals through pairs of opposite vertices. For comparison, an octahedron appears on the right side.

Let's describe that set of symmetries as a group of shuffles applied to the four diagonals. Begin with the rotations that return the cube or octahedron to its original position as a whole rather than point for point. There are three types of special location on either polyhedron: vertices, mid-edges (midpoint of an edge), and face-centers. Indeed, the specialized nature of these points is due to the fact that, for each one, some rotational symmetry fixes it. Under a symmetry, a special point of a certain kind has to move to a special point of the same kind. As a consequence, any axis of rotational symmetry passes through an antipodal pair of special points. In addition to the trivial transformation (which belongs to every group), Table 5.1 lists a cube's three types of axis, the respective order of turning about each axis, how many axes there are of a certain type, and the total number of distinct rotations corresponding to a type of axis. Note that $m - 1$ is the number of rotations about an m-fold axis that are not trivial.

Table 5.1 Rotational Symmetries of the Cube and Octahedron.

Axis-type (cube)	Order m-fold	# Axes a	# Rotations $(m - 1) \cdot a$
trivial	–	–	1
mid-edge	2-fold	6	$6 = (2 - 1) \cdot 6$
vertex	3-fold	4	$8 = (3 - 1) \cdot 4$
face-center	4-fold	3	$9 = (4 - 1) \cdot 3$
total	–	–	24

That the cube's family of rotational symmetries and the shuffles of four objects each add up to 24 is tantalizing. However, before declaring that the two groups are equivalent, we need to know that they are structurally the same. To establish this likeness, work out the correspondence between rotations and shuffles of the diagonals. Refer to the cube in Figure 5.6 and give the sphere a quarter-turn counter clockwise about the vertical axis (viewed from above the north pole). Call this transformation F (for face axis). Invoking the first symmetry principle, successive applications of F are also symmetries and, so, shuffle the diagonals:

$$F: R \to Y \to B \to G \qquad F^2: R \leftrightarrow B \quad Y \leftrightarrow G \qquad F^3: R \to G \to B \to Y$$

The other two axes between face-centers produce similar shuffles of vertex-pairs. A one-third turn V and two-thirds turn V^2 about a vertex-axis—for instance, along the red vertex-pair—cycles the other three diagonals:

$$V: B \to G \to Y \qquad V^2: B \to Y \to G$$

The remaining type of rotational symmetry gives the spherical cube a half-turn E about an axis through antipodal mid-edges—between red and blue vertices, say. Such a motion rearranges the diagonals:

$$E: R \leftrightarrow B$$

Checking that rearrangements of vertex-pairs (or diagonals) produced by the cube's rotational symmetry group are distinct elements of the complete shuffle group S_4 is a simple matter. Since there are 24 such shuffles, the two groups are equivalent and so, we have a one-dimensional space that conforms to Klein's Rule for the quartic polynomial.

$n = 5$

For the first time, we encounter the circumstance in which a complete shuffle group—S_5 in this case—fails to admit description as a group of rotations acting on the sphere. In light of this significant geometric fact, we examine the even shuffle group A_5, which leads to an iconic structure exploited by Klein's method of solving the quintic. As pictured in Figure 5.7, the *icosahedron* consists of twenty triangular faces joined together five per vertex [32]. Its companion polyhedron is the twelve-faced *dodecahedron*. The graphic also illustrates how five tetrahedra—triangle-based pyramids—can be inscribed in either framework.

Having added the tetrahedron, we can take note of a grand fact of polyhedral geometry. First, let's define terms. A regular polygon is equilateral and equiangular. In a *regular polyhedron* 1) the faces are regular polygons of the same size and shape and 2) the same number of faces surround each vertex. Now for the extraordinary claim: The tetrahedron, cube, octahedron, dodecahedron, and icosahedron are the only regular polyhedra. Also known as Platonic solids—due to Plato's having built a cosmology around their properties[40], they can be thought of as embodying maximum symmetry in the sense that their faces are indistinguishable as are their vertices.

Returning to a symmetrically-inscribed tetrahedron, notice that its vertices land on four of the icosahedron's twenty face-centers (or dodecahedron's vertices). The tetrahedra provide five objects—distinguished by colors R, B, G, Y, and C—that rotational symmetries of the icosahedron shuffle.

Actually, there's a bit more going on here. Extending any edge of either polyhedron around the entire sphere produces a circle that serves as a "line" of reflective symmetry relative to the system of edges. Although we aren't using these reflections for our purposes, they make for an interesting effect when applied to the tetrahedra; namely, after reflection, a different collection of five tetrahedra appear. In the new system, five sets of four vertices once again exhaust the icosahedron's face-centers. The property responsible for this phenomenon is known as *chirality* or handedness. A figure on a sphere is *chiral* if, using only rotations, it cannot be made to coincide with its mirror image. Being reflectively symmetric, an individual tetrahedron is achiral, but when configured with its four associates, the aggregate becomes chiral. You can glimpse a system's handedness in the pentagonal "fan-blade" centered at a point of five-fold symmetry, where the five blades are parallelograms.

Characterizing the icosahedral symmetries in the same fashion as in the case of a cube, Table 5.2 counts rotations by classifying their types of axis.

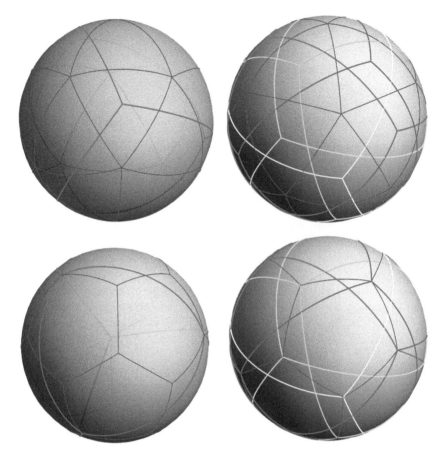

Figure 5.7 Icosahedron and dodecahedron. The top row displays the icosahedral configuration while the dodecahedral pattern appears at bottom. Each structure is formed from 30 (black) edges of which five coincide at a vertex of the icosahedron and three join together at a vertex of the dodecahedron. Also depicted here are inscribed tetrahedra. On the left side, only one such four-faced pyramid is evident on a slightly transparent sphere. Tetrahedral vertices are situated at four of the icosahedron's face-centers. At right, you see opaque spheres with five tetrahedra elegantly knit into the icosahedral/dodecahedral forms so that tetrahedral vertices exhaust the 20 face-centers/vertices. With these illustrations you can clearly perceive the manner in which the icosahedron's symmetries act on the tetrahedra. For instance, take the arrangement at top right and an axis through the vertex near the image's center. A counter-clockwise one-fifth turn (five-fold symmetry) moves the tetrahedra as indicated in Table 5.2.

As before, the axes pass through antipodal pairs of special points: 30 mid-edges, 20 face-centers, and 12 vertices. The table also shows what a shuffling of tetrahedra looks like for an axis-type. Now, use these shuffles to rearrange the terms of the square-root of the quintic's discriminant $\sqrt{D_5}$—as explained in Section 4.3. You will then verify that each shuffle returns $\sqrt{D_5}$ unchanged and so, belongs to the group of 60 even shuffles A_5. Every rotational symmetry is thereby matched with an even shuffle. Therefore, the two groups turn out to be equivalent and a spherical icosahedron (or its rendering on a complex line) gives us a one-dimensional space that complies with Klein's Rule.

Table 5.2 Rotational Symmetries of the Dodecahedron and Icosahedron.

Axis-type (icosahedron)	Order m-fold	# Axes a	# Rotations $(m-1) \cdot a$	Representative shuffle
trivial	–	–	1	
mid-edge	2-fold	15	$15 = (2-1) \cdot 15$	R ↔ B C ↔ G
face-center	3-fold	10	$20 = (3-1) \cdot 10$	R → C → G
vertex	5-fold	6	$24 = (5-1) \cdot 6$	R → B → G → C → Y
total			60	

$n = 6$

Taking on the problem set by Klein's Rule when there are six roots leads to another watershed. Neither S_6 nor A_6 can be captured as groups of transformations acting on the complex sphere or line [1]. Consequently, we look to higher dimensional space. Late nineteenth and early twentieth century mathematics spent considerable effort systematically searching for spaces on which groups of various kinds can be expressed in terms of transformations. This activity gave impulse to deep theoretical developments including group representations and invariants [14]. A *representation* of a group G on a space X is a set of transformations on X whose collective structure is, to some degree, in agreement with that of G. Satisfying Klein's Rule requires the representative group of transformations to be exactly equivalent to the represented group.

A phenomenon more general than that of a representation is when a group *acts* on a set X whereby each group element corresponds to a particular rearrangement of the set's members. For an example, take the manner in which the cube's group of rotational symmetries shuffles its set of vertices. Given a point p in X, if you gather the points generated by applying all of the shuffles affiliated with the acting group, you get a subset of X called the *group orbit* (or just orbit when usage is unambiguous) of p. Orbits of this type play an important role in our equation-solving methodology.

Among the fruit born by these pursuits are three cases pertinent to the Kleinian program for computing roots of sixth and seventh-degree polynomials. Discovered by Valentiner and advanced further in subsequent work, the collection of 360 even shuffles A_6 is equivalent to a group of transformations—the Valentiner group—that act on a space with two complex (and four real) dimensions [47, 22, 48]. You can think of this space as a kind of four-dimensional polyhedron called a *polytope*—a higher-dimensional surface with special points, curves, and surfaces sculpted into it. The complete shuffle group S_6 cannot be described as such. Keeping with terminological preferences, call this setting a *complex plane*.[1] By way of caricature, imagine a standard co-ordinate setup with two axes each of which is a complex line. A location is specified either by a pair of complex numbers or a quadruple of real ones.

The algebraic and geometric structures erected by Valentiner's group are quite rich [7]. Here, attention focuses on a basic question. As the Valentiner transformations act, what six objects experience even shuffling—that is, when such a shuffle is converted into rearrangements of roots, the square-root of the discriminant $\sqrt{D_6}$ does not change sign? Working in A_6, say that you possess the six shuffled objects and pick any one of them. The collection of shuffles that fix the object in place forms a subgroup whose members move the remaining five objects in the same way that elements in A_5 rearrange five things. In other words, the subgroup and A_5 are equivalent.

On this landscape expressed by algebra is where geometry rises. Each of the A_5-equivalent subgroups fixes a complex version of an ellipse known as a conic section (or just conic); this means that the set is unchanged, not the individual points that belong to it. Described in real dimensions, such a conic is a sphere-like surface sitting inside the four-dimensional Valentiner polytope. By virtue of its A_5 symmetry, each conic is configured as an icosahedron. The resulting system of six icosahedral spheres supplies the objects required by our basic question.

What this picture calls to mind immediately is the set of five A_4-symmetric tetrahedra that are coaxed out of the icosahedral transformations as they act on a sphere. In light of this comparison, there's a bit more to be said. Recall that a second assemblage of five tetrahedra resides on the icosahedron's surface. Each set is a mirror image of the other. Analogously, another system of six icosahedral conics occurs under the operation of Valentiner's group. This outcome derives from a peculiar property associated with the structure of A_6. The quirk stems from the fact that the group of icosahedral rotations has a description in terms of shuffles applied to six objects in addition to five. Here's a brief description of how this extra bit of structure happens to exist. Using the same technique that we applied to the cube's four diagonals, label the six pairs of antipodal icosahedral vertices 1 to 6 and track their movements as the icosahedron rotates symmetrically. There are six distinct ways of labeling the vertices each of which produces an A_5-equivalent subgroup of A_6. For each of these subgroups we get an icosahedral conic and, in sum, a second system of six objects shuffled by Valentiner transformations.

---

[1]Caution: as mentioned early on, the term 'complex plane' is commonly used to refer to what we call a complex line.

An example of the sort of elegant geometry that motivates our approach to equation-solving shows up when the conics in the two systems intersect. The diagrams in Figure 5.8 use two real dimensions to depict the configurations figuratively. Take one conic C from either system and regard it as a spherical icosahedron. When another conic in the same system crosses the one selected, it does so in four points that form one of the five tetrahedra inscribed in C. Intersecting C with the five other conics in its system gives all twenty icosahedral face-centers. By taking the points that C has in common with the six conics in the complementary system, each intersection contributes two points that happen to be one of C's six antipodal vertex-pairs. Two spherical surfaces that intersect at two or four locations presents a visual challenge. So it goes with curiosities found in higher-dimensional spaces whose coordinate description uses complex numbers.

There are 45 elements in A_6 that exchange two pairs of objects. For instance, 1 and 2 trade places as do 3 and 4 while 5 and 6 stay put. Every member of A_6 can be realized as some combination of these double-swap shuffles. In Valentiner's group, such a shuffle corresponds to a *complex reflection*, referring to a transformation on two-dimensional complex space that fixes every point on a one-dimensional complex mirror. As we'll see, these 45 lines of reflective symmetry form an intricate web within the Valentiner structure while playing an important dynamical role.

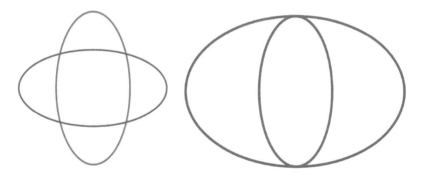

Figure 5.8 Cartoon illustration of intersecting icosahedra. Conics in the same system cross at four tetrahedral points (left). Conics in different systems meet at a pair of antipodal vertices on each icosahedron (right).

$n = 7$

When it comes to shuffling seven objects, Klein's influence is felt again. Advances in nineteenth-century algebra established the futility in attempting to represent A_7 on the same two-dimensional stage where the Valentiner group plays. Such a representation *is* available by moving up a dimension to a space expressed by triples of complex numbers [31]. There's still much to learn about this group's algebraic and geometric properties.

Giving priority to aesthetic sensibility, we now deviate from interrogating only the groups A_n and S_n. Investigations into the symmetries defined on a certain type of surface led Klein to recognize the existence of a remarkable group [22]. Collectively its 168 elements are structurally equivalent to a subgroup of A_7 and individually transform the same space as the Valentiner "motions." What's more, Klein's group constitutes the symmetries of a polytope bearing similarities to the one carved out in the Valentiner setting [9]. The principal resemblance between the two formations is evident in their respective systems of conics. Housed in the Klein polytope are two sets of seven conics, each of which has the structure of a cube. Noting that the cube's symmetry group holds 24 rotations, the key numerical relationship here is $168 = 7 \cdot 24$. As in the Valentiner situation, intersections among Kleinian conics pick out special sets of points. The cubes associated with Klein's group produce seven objects—in two ways—that are shuffled in the manner prescribed by the relevant members of A_7. As such, the group provides a tool for solving heptic (seventh-degree) equations whose symmetries it exemplifies.

$n > 7$

At this point, we reach a juncture. Excepting the simplest case of S_2, all of the groups considered so far have special representations. To clarify the meaning of 'special,' notice that, for any complete shuffle group S_n, every member can be captured by a transformation in n-dimensional space. Given a point

$$(z_1, z_2, \ldots z_n),$$

let an element of S_n act by shuffling the point's coordinates. For instance, the shuffle

$$1 \to 3 \to 4$$

in S_4 transforms the point (z_1, z_2, z_3, z_4) into (z_3, z_2, z_4, z_1). Obviously, the action amounts to shuffling the subscripts. Call this way of realizing a complete shuffle group in n-dimensional space the *standard representation*.

It turns out that we can lower the dimension associated with S_n. The collection of points whose coordinates sum to zero is an $(n-1)$-dimensional space X_n. In other words, points in X_n satisfy a condition:

$$z_1 + z_2 + \cdots + z_n = 0.$$

The root function on the left-hand side is unchanged when the coordinates rearrange. Known as a hyperplane, X_n is a generalization of a line, plane, and space. You're free to select $n - 1$ coordinates of such a point $(z_1, \ldots z_n)$ while the nth is determined. Continuing with S_4 as an example, you can express points in X_4 with three free variables:

$$(z_1, z_2, z_3, z_4) \qquad z_4 = -(z_1 + z_2 + z_3).$$

This observation makes it easy to identify a group orbit consisting of n objects when S_n acts on X_n. Using $n = 4$ again, we have a set of points:

$$(-3, 1, 1, 1), \ (1, -3, 1, 1), \ (1, 1, -3, 1), \ (1, 1, 1, -3).$$

Affiliated with each of these points is a hyperplane whose dimension is one less than that of the containing space X_n, meaning these four hyperplanes are two-dimensional and so, are planes in the complex sense.

What's more, there's a means of projecting away another dimension of X_n so that we arrive at an $(n-2)$-dimensional space. We can safely ignore the details involved in this operation, notwithstanding the fact that every case listed in Table 5.3 also employs this projection. To summarize, above the double line group representations are special—non-standard, while the only type of representation for groups listed below the double line are standard ones.

Table 5.3 Following Klein's Rule. In ascending order, the table lists basic facts pertaining to a symmetry group that's equivalent to the relevant shuffle group. The dimension where a representation acts is as small as possible. Mentioned here is a dihedron. Thinking of it as a sphere with a single vertex on an edge that coincides with the equatorial circle, there is but one rotational symmetry.

Group	Dimension	Symmetric structure	Shuffled objects
S_2	1	dihedron	2 points
S_3	1	double triangular pyramid	3 points
S_4	1	cube/octahedron	4 antipodal vertex-pairs
A_5	1	icosahedron/doecahedron	5 tetrahedra (4-point sets)
A_6	2	Valentiner polytope	6 icosahedra (conics)
A_7	3	unknown	unknown
A_n or S_n	$n-2$		n points/hyperplanes

Representations of even or complete shuffle groups serve as algebraic and geometric platforms upon which we construct a program for solving the associated equations. Such actions furnish a setting in which a polynomial's symmetry is realized. Now the burden is to devise a mechanism that breaks the symmetry found there, as required by the Galois Doctrine. To achieve the desired outcome, we use dynamical processes. Before delving into the way this works, let's fill in some mandatory background.

5.3 FUNDAMENTALS OF DYNAMICS

Much of math and science involves the exploration of systems governed by axioms or rules of some kind. Frequently, the governing principles determine a system's state at any given time. The notion of time here need not indicate literal chronology, but refers to a quantity that tells us how states are ordered. Of particular interest is the evolution of a system's state: given an exact specification of a state at some time, how do subsequent states unfold? We call such an entity a *deterministic dynamical system* in contrast to one whose states are subject to random influences. Systems can be reversible so that you can interrogate the states that precede as well as follow the current one.

Dynamical systems come in two flavors, one of which is a *continuous* system given by a solution to a set of differential equations. Mathematical models of this sort are ubiquitous in physics, engineering, and economics. The simplest case of such a model is a single equation such as

$$\dot{x} = 0.$$

Its solution calls for a function $x = f(t)$ that satisfies the condition expressed by the equation. Think of $x = f(t)$ as a function of a time variable t and interpret $\dot{x}$ as the instantaneous rate at which x is changing with respect to time. In the language of calculus, $\dot{x}$ is the time derivative of x. The equation states that x is experiencing no change, and so the solution is a function whose value is constant:

$$x = f(t) = c \qquad \text{for any value of } t.$$

Giving c a specific value requires another condition such as $f(0) = 1$.

Now, in the original equation, replace 0 on the right side with 1 and the solution has the form

$$x = f(t) = t + c.$$

Giving c a definite value, the solution is a trajectory that "flows" along an x-axis at a constant rate of one unit of distance per unit of time. If x represents position on the axis, $\dot{x}$ expresses velocity. Since it amounts to a continuous process, such flowing behavior justifies the terminology.

Our concern is with the other type of dynamical system. Such a process is *iterative* and evolves in *discrete* "time" steps. The discussion here treats only those concepts and pieces of theory applicable to root-finding.[2] At the core of discrete dynamics is a function-like appliance called a *map* with a specific property. Generally, a map f from a domain set Z to a codomain W associates a given point z in Z with some point w in W. Put succinctly, given an element z in Z, $w = f(z)$ is a member of W.

Invoking a simple illustration, let

$$w = f(z) = \frac{1}{z^2} \qquad \text{alternatively} \qquad z \xrightarrow{f} \frac{1}{z^2} \quad (z \text{ "maps to" } 1/z^2)$$

with Z and W the points that are respectively inside and outside the circle whose center is zero and radius is one. This example conforms to the specific type of map that we will deploy when engaged in solving equations, one that's expressed as a ratio of polynomials $P(z)$ and $Q(z)$. A *rational map* has a form

$$z \xrightarrow{f} \frac{P(z)}{Q(z)}$$

---

[2]See [15] for a more comprehensive treatment.

analogous to that of rational numbers—ratios of integers. Its degree is the highest power of the variable that appears in either the numerator or denominator.

Returning to the example, f has a degree of two. Now, take a point z in Z within the unit circle, so that the length $|z|$ is less than one. Then the length

$$|z^2| = |z|^2$$

is also less than one and the corresponding point w has length

$$|w| = \left|\frac{1}{z^2}\right| = \frac{1}{|z^2|} = \frac{1}{|z|^2}$$

which is greater than one. The upshot from these observations is that f maps Z to W—the unit circle's interior to its exterior. But, since W is not part of f's domain, you cannot iterate f—that is, apply f to w. This impediment is easily avoided by electing to use the entire complex line as the domain Z. There's even room for a point at infinity ∞ as well, since we can ascribe sensible behavior to the map there:

$$f(0) = \infty \quad \text{and} \quad f(\infty) = 0.$$

In this domain, a point z whose length is greater than one yields a w with a length less than one.

In general, when $W = Z$, meaning that the domain and codomain are the same set, f becomes a complex dynamical system on Z by means of iteration. For the maps that we consider, the set Z is a complex space or a subset thereof. To see how iteration goes, take an element z_0 in Z. Since $z_1 = f(z_0)$ belongs to Z, it can be fed back into f, producing a point in Z that we describe with an abbreviation

$$z_2 = f(z_1) = f(f(z_0)) = f^2(z_0).$$

The notation f^2 indicates that the map is applied twice. In turn, f treats the output $f^2(z_0)$ as a new input and the next point in the sequence is

$$z_3 = f(f^2(z_0)) = f^3(z_0),$$

and so on. Continuing this recursive process generates the basic object of study in dynamics: an infinite sequence of values that begins with a point z_0 and grows by iterating f on this *initial condition*. Designate such a progression the *orbit* or trajectory of z_0. When needed for clarity, call it the f-orbit. Written out in two ways, the orbit is an infinitely long ordered list:

$$\text{orbit of } z_0 = (z_0,\ z_1,\ z_2,\ \ldots) = (z_0,\ f(z_0),\ f^2(z_0),\ \ldots).$$

For the sample map discussed above, the orbit of $z_0 = 2/3$ is

$$\left(\frac{2}{3},\ \left(\frac{3}{2}\right)^2,\ \left(\frac{2}{3}\right)^4,\ \left(\frac{3}{2}\right)^8,\ \ldots\right).$$

Notice that 'orbit' is doing double duty as we also use it in the context of a group action. The two concepts are related, but should be distinguished. When necessary, we refer to a distinct type as either a group or dynamical orbit. Map iteration produces a deterministic orbit sequence, meaning that no random events affect the outcome. We now examine and later utilize dynamical systems that are deterministic and discrete. In Part II, we have occasion to consider non-deterministic dynamics.

The orbit concept is foundational for dynamics and supports the subject's basic question—one that's easy to state, but often difficult to resolve.

> In discrete complex dynamics, the **fundamental problem** is to work out the long-term behavior of an orbit. That is, what happens to $f^k(z_0)$ as k becomes arbitrarily large?

Analyzing the example orbit just given is a simple matter. Values in odd positions such as $2/3$ and $(2/3)^3$ involve increasing powers of $2/3$. For a number less than one, its powers tend to zero. Terms occupying even positions are enlarging powers of $3/2$; quantities of this sort grow arbitrarily large. Accordingly, the orbit's tendency is toward the repeating sequence

$$(0, \ \infty, \ 0, \ \infty, \dots).$$

What makes the treatment of this orbit easy to discern is the fact that z_0 is a real number.

Let's muddle the issue by "complexifying" the initial condition:

$$\left(\frac{2}{3} + i, \ -\frac{45}{169} - \frac{108}{169}i, \ \frac{119}{81} - \frac{40}{27}i, \ -\frac{1568079}{815730721} - \frac{187382160}{815730721}i, \ \dots \right).$$

Expressing the orbit values as decimals and computing several more iterates will likely convince you that the tendency of this orbit agrees with the preceding one. However, there's a more potent way to think about the problem. Rather than watching a single orbit, we can follow the behavior of an infinite collection of orbits by creating a "companion orbit" whose entries are lengths of the complex numbers in the original list. For the starting point, we have

$$|z_0| = \left| \frac{2}{3} + i \right| = \frac{\sqrt{13}}{3}.$$

The next item in the orbit of lengths is

$$|z_1| = \left(\frac{1}{|z_0|} \right)^2 = \left(\frac{3}{\sqrt{13}} \right)^2.$$

At each step, obtain the subsequent length by squaring the reciprocal of the one that goes before. What you get is

$$|z_2| = \left(\frac{1}{|z_1|} \right)^2 = \left(\frac{\sqrt{13}}{3} \right)^4 \quad \text{and} \quad |z_3| = \left(\frac{1}{|z_2|} \right)^2 = \left(\frac{3}{\sqrt{13}} \right)^8.$$

Note that this "length orbit" is not a dynamical orbit in the true sense; we aren't iterating the length function. Rather, we're looking for a useful property of the orbit whose behavior we want to understand. And we find one in the sequence's tendency, a situation that's basically the same as it was when $z_0 = 2/3$. Powers of $\sqrt{13}/3$ head toward infinity, while raising $3/\sqrt{13}$ to an arbitrarily high power pushes the values to zero. Here, the big difference is that the condition $|z_0| = \sqrt{13}/3$ applies to all complex numbers on the circle whose radius is $\sqrt{13}/3$ and center is 0.

Adopting our alternative perspective allows us to see that so long as the length of z_0 is not equal to one, the oscillating and asymptotic behavior present in the examples will persist. Doing so amounts to a description of f's *global dynamics*, meaning a classification of all orbits. A complete account would treat the unit circle, each element of which has a length of one. What happens there? For a short answer, the circle is a world where chaotic activity occurs. One of the many ways to characterize "topological chaos" is 1) transitivity: some point has an orbit that gets arbitrarily close to every spot on a specific set and 2) sensitivity: given a prescribed distance, there exist two arbitrarily close initial conditions whose orbits eventually become separated by the designated amount. Remember, an orbit is infinitely long. Although the maps that we employ for equation-solving purposes behave chaotically in places, the focus is on parts of space where chaos is absent. Complementing the exhaustive discussion held here is one that addresses a map's *local dynamics*, to which we now turn.

An especially simple type of dynamical orbit is one that repeats. Beginning with z_0, the orbit returns to that point after some number of iterations m. Put symbolically, $f^m(z_0) = z_0$. When no smaller number of steps lands on z_0, call that point *periodic* of period m and say that its orbit is a periodic cycle (or just cycle) of length m. The minimal case occurs when $m = 1$ and a point is *fixed*. Importantly, you can view a point whose period is greater than one as a fixed point relative to the mth iterate map f^m. As a consequence, understanding how a map acts near a fixed point suffices for grasping its behavior along a cycle. The diagram in Figure 5.9 sketches how the iterative process works for a cycle of length m. The circle centered at z_0 indicates a *neighborhood* of points relatively close to z_0. If this neighborhood is very small, the circles whose centers are points in the cycle represent approximations—not to be taken literally—to the actual outcome of applying iterates of f to points near z_0. The smaller the neighborhood of z_0, the closer the diagram is to capturing what actually happens. Assessing the map's local dynamics across the entire cycle reduces to the examination of how f^m acts on z_0's neighborhood. Before discussing the issue, we require another rare kind of point the characterization of which likewise depends on how the map affects a proximate region.

A point c is *critical* when f "wraps" a neighborhood of c around the location $f(c)$. Regarding the neighborhood as a region on an elastic sheet provides a physical metaphor. The description here applies to a map that runs on a one-dimensional space. In higher dimension, the activity is more elaborate,

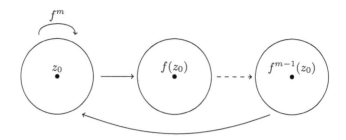

Figure 5.9 Local dynamics along a periodic cycle. Points inside a circle map roughly to the next circle—ignoring their size.

but the basic ideas carry over. Position yourself at c and, as Figure 5.10 portrays, watch what happens to the points on a nearby arc of a circle that forms an angle at c. Proximity here refers to an infinitely small or infinitesimal distance R between c and the arc—interpreted as a quantity that shrinks to nothing without actually reaching zero. Applying f, you arrive at your destination $f(c)$ and observe the transformed arc to be much closer—infinitely so—to $f(c)$ than the original arc was to c. The relative distance between c and its associated arc is infinitely larger than the distance between $f(c)$ and its associated arc (taken to be R^2 here). That is to say, the map crushes an infinitesimal neighborhood of c down to something that resembles a point. Consider the ratio of respective distances:

$$\frac{R}{R^2} = \frac{1}{R},$$

which is an infinitely large value. In addition, the angle created by the arc has increased. Exemplifying a hallmark of criticality, the sheet shrinks radially and stretches in the angular direction that winds around the point—call this behavior "shrink-wrapping."

We can now introduce the dynamical character that plays a crucial part in our story. Assume that f is a map with a fixed point a. If there is a neighborhood of a in which every element has an orbit that asymptotically approaches a, we say that a is *attracting*. A natural object associated with an attracting fixed point is the *largest* neighborhood every member of which has an orbit that ultimately homes in on a. Call this maximal set B_0, the *immediate basin of attraction*. For topological reasons, every pair of points in a neighborhood can be connected by a (possibly curved) segment entirely contained in that set. The idea is that no point on the boundary of B_0 has an orbit attracted to a so that the set can't be expanded while retaining its defining property. Typically, there are other regions—separated from B_0— that collectively map on top of it. The orbit of any point in such a collection also gets arbitrarily close to a. Figure 5.11 indicates how to push the iteration backward any number of times: for each set A that maps over B_0, there are sets that map on top of A, and so on. Gathering all the points in these sets

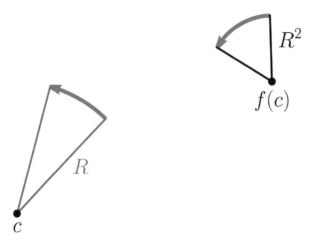

Figure 5.10 Critical behavior. Points on the magenta arc about a map f's critical point c are an infinitely small distance R from that location. About $f(c)$ is a magenta arc where f sends the arc about c, while the distance from respective points shrinks by an infinite factor. The other feature that signals the presence of a critical point involves the angles formed by the respective arcs and points. Specifically, the angle formed by the black segments at $f(c)$ is some integer factor larger than the angle created by the blue segments at c (it's two in this example).

produces the complete basin of attraction (alternatively called an attracting basin or just basin) B consisting of all points whose orbits tend to a. Running the map in reverse like this can lead to the discovery of baroque fractal shapes that, as we will see, emerge from visual displays of a map's basins.

Being analogous to fixed points, periodic cycles can also exhibit attracting behavior. According to previous commentary, such a condition can be analyzed using an appropriate iterate of the map. Returning to the example $f(z) = 1/z^2$, we identified 0 and ∞ as a period-2 cycle whereby the second iterate $f^2(z) = z^4$ fixes the two points. Deploying f^2, each member of the cycle attracts points residing in either the unit circle's interior as the basin of 0 or exterior as the basin of ∞. Acting as neutral territory, the circle maps onto itself in a chaotic fashion. We describe the global dynamics by noting that the basin of the cycle $(0, \infty)$ consists of the regions inside and outside the unit circle. Among rational maps, this one is highly unusual in that the attracting cycle's immediate basin is the same as its overall basin. Hence, we find no new pieces of the basin when iterating in reverse, and so, the basin fails to exhibit a fractal nature.

Apply this line of thought to a garden variety attracting m-cycle. Over the cycle's course, each point consumes a collection of nearby points. The implication is a one-in-all-in rule: if an orbit closes in on one member of the cycle, it will do so for all members.

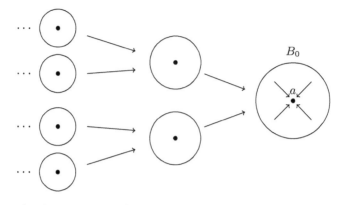

Figure 5.11 A schematic view of a basin. The attracting fixed point a resides in its immediate basin B_0. Indicated by arrows, two sets each map over B_0 and are themselves mapped over by two other regions. Distinguished points in these sets eventually map to a. The cascade continues as an iteration in reversed time. For this illustration, the map has degree two, which accounts for its two-to-one behavior.

Our attention concentrates on one-dimensional systems, where two types of attracting fixed point occur—to be examined presently. Moving to higher dimension, attraction is more intricate, but you can glean the gist of how things work by splitting up the dynamics near a fixed point into lower dimensions.

As already noted, the phenomenon of attraction has both local and global qualities. To say a bit more about the former, let's examine its asymptotic nature in greater detail. Analyzing dynamics locally means thinking about what a map "looks like" near an attracting fixed point. An important technique in dynamical theory and practice—one that also plays a key role in equation-solving—involves changing coordinates. We now illustrate the process in two ways corresponding to the types of attraction. Our examples are taken from the intensely-studied family of *logistic* maps each of which is rational with degree two. (In fact, logistic maps are expressed by quadratic polynomials—a special kind of rational map.)

In the first instance, take

$$z \xrightarrow{f} \frac{3}{2}z(1 - z) = \frac{3z - 3z^2}{2}$$

defined on the entire line of complex numbers. Solve the condition $f(z) = z$ to obtain fixed points at 0 and $1/3$. To ascertain what happens when z is near $1/3$, change to a new coordinate w that's close to 0 by "sending" $1/3$ to 0. (There are useful techniques from calculus—not assumed here—that can be brought to bear on this question.) You can think of this change as a renaming of streets and renumbering of houses. When expressing the original map in terms of the new coordinate, it takes the form

$$w \xrightarrow{F} \frac{1}{2}w$$

where we've introduced a new map name F, one that's different from yet affiliated with f. (For readers with a calculus background, the factor of $1/2$ is the derivative of f at $x = 1/3$.) The great thing about this coordinate transformation is that the dynamics near 0 in "w-space" is easy to work out and then invert in order to recover what's going on in the original space described by z. If you iterate f symbolically, the result is an algebraic morass whereas, an orbit produced by F is simple to state:

$$\left(w_0, \frac{1}{2}w_0, \left(\frac{1}{2}\right)^2 w_0, \ldots, \left(\frac{1}{2}\right)^k w_0 \ldots \right).$$

The long-term behavior is governed by the factors $(1/2)^k$, regardless of the initial point w_0's location (although w_0 is restricted to values close to 0). Raising $1/2$ to arbitrarily high powers generates values that dwindle to nothing, meaning that 0 is an attracting fixed point with respect to F. Now, reverse the coordinate change, passing from w to z. You see that the dynamical behavior near $z = 1/3$ under f resembles F's activity around 0.

As for the other fixed point of f, namely 0, we can again apply a coordinate change near $z = 0$ whereby the transformed map is

$$w \xrightarrow{F} \frac{3}{2}w.$$

Here, an orbit's tendency is determined by the factor $(3/2)^k$, which increases in value as k grows. No matter where an orbit begins in the vicinity of 0, it moves away. Such a point is *repelling* and remains so when carrying the dynamics back to the original setting of the map f.

A second kind of attraction results when a critical point has a periodic orbit. We soon will learn that the global dynamics of a one-dimensional map whose critical points are periodic is quite robust. Although this property is not shared exactly by maps in higher dimension, they can exhibit analogous behavior. Maps of this sort lie at the heart of routines that solve equations.

For a sample map, let's use another member of the logistic family:

$$z \xrightarrow{g} 2z(1 - z) = 2z - 2z^2$$

noting that 0 and $1/2$ are fixed. Once again, 0 turns out to be repelling. The point $1/2$ is critical and when you change from z to w coordinates by assigning $1/2$ to 0, the result is

$$w \xrightarrow{G} w^2$$

when w is infinitely close to 0.

Figure 5.12 compares the two modes of attraction by sketching how each map acts on a circle with suitably small radius R centered at a fixed point a. As noted, these images portray an *approximation* of what a map f or g actually does to points near the attracting site.

When a is not critical, the associated map is

$$w \xrightarrow{F} sw \qquad |s| \text{ less than } 1$$

so that the outer circle maps to the middle one, which f sends to the inner circle. Where one circle maps onto another, the coefficient length $|s|$ is a shrink factor between the two. As a ratio of radii,

$$|s| = \frac{\text{middle radius}}{\text{outer radius}} = \frac{|s|R}{R} = \frac{\text{inner radius}}{\text{middle radius}} = \frac{|s|^2 R}{|s|R}.$$

In our example, $s = 1/2$. The other feature of a map's behavior near a concerns its effect relative to angle measure. The diagram renders the circles as a curved arrow with a "tail" mapping to a tail and a "tip" to a tip. So, points on the outer circle go to the one in the middle by turning through a fixed angle between the red and green radial segments. The turning angle is the same when mapping from middle to inner circle—between green and blue radii. In a nutshell, the full circular arrow ($360°$) experiences a transformation that both shrinks and turns it.

In the event that a is critical, g looks like $w \to w^2$ and so, collapses a circle whose radius is an infinitesimal R onto one with a radius of R^2. Taking the ratio of the two radii gives a factor of shrinking equal to

$$\frac{\text{inner radius}}{\text{outer radius}} = \frac{R^2}{R} = R.$$

Comparatively, the inner circle is infinitely smaller than the outer one, implying that an orbit approaches a fixed critical point very rapidly. We call such a point *superattracting*. To manifest g's activity in an angular (or polar) sense, the outer circle is split into two semi-circular arrows. Each of the semi-circles maps onto the whole inner circle so that the radius-R circle wraps *twice* around the circle of radius R^2. When a map resembles the power function

$$w \to w^k \qquad k \text{ greater than } 2$$

close to a fixed critical point, it shrink-wraps a neighborhood of the superattracting point to a degree greater than two. Furthermore, the big circle winds around the small one k times.

A map has *reliable dynamics* when there are attracting cycles whose basins contain any randomly chosen point. This is a statistical statement and can be framed metaphorically. Think of the complex number line as a dart board. (Recall that it forms a plane in the usual sense.) When thrown in a random fashion, there's zero chance that a dart will miss landing on a basin. This outcome does not imply that all points are in basins. Indeed, there can be and usually are an infinite number that aren't. What matters is how densely the non-basin points are distributed. Maps with reliable dynamics are central to the root-harvesting algorithms to come. A seminal theoretical result on orbits of critical points guides the discovery of reliable one-dimensional maps.

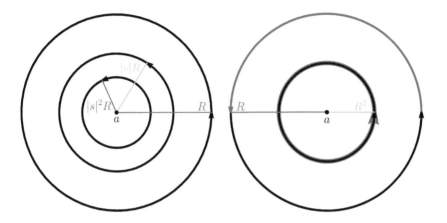

Figure 5.12 Two models of attraction. At a non-critical fixed point a, any small neighborhood shrinks by a constant factor $|s|$ (left). A neighborhood of a fixed critical point squashes down by a factor that's relatively larger when the points are closer to a (right).

> Suppose that *all* of a rational map's critical points are periodic, meaning that every critical point belongs to a superattracting cycle. According to the **Critical Theorem**, the basins of superattraction fill up the complex line in the sense of reliable dynamics. That is, the orbit of a random starting point will asymptotically approach one of the superattracting cycles.[36, Theorem 1.3]

While still important in higher dimensions, the behavior of critical points does not control global dynamics to the extent seen on a complex line. The principal difference between criticality in one and several dimensions is that in the former setting, there are a finite number of critical points—limited by a map's degree—whereas, in the latter, an infinite number are present.

Let's now conclude this abbreviated tour of relevant dynamical results that emerge when rational maps undergo iteration. Pulling together ideas and tools previously discussed, we undertake a case study that aspires to resolve the fundamental problem at a global level.

Say that you're handed the map

$$f(z) = \frac{z^2 + 1}{2z}$$

and tasked with classifying the orbits of all points on the complex line. Figure 5.13 illustrates a narrative through which we discover a solution. At top left is a space described by the coordinate z. Below this "z-space" is a downward vertical arrow which captures the main question. At the start, we don't know how to classify orbits. In order to reach the state of affairs indicated at bottom left, travel clockwise through the diagram, beginning at top left.

The initial step is to calculate the fixed points, which happen to be -1 and 1. Using calculus techniques, you can readily find that each of these points is also critical. Alternatively, use a non-calculus approach to determine whether or not the fixed points are attracting. Namely, introduce a new coordinate w by sending one fixed point, say 1, to 0 and the other to ∞. The top horizontal arrow defines the transformation from z to w. Though it's not so important to know how to derive the particular formula (it appears here for completeness), what is crucial to take in is that the change of coordinates transforms the vertical line in z-space into the unit circle in w-space where we obtain a map with the form

$$F(w) = w^2.$$

This power map made a previous appearance in the context of critical points. However, then its domain was restricted to an infinitesimal neighborhood of a fixed point. In the current situation, the coordinate change is universal—that is, valid for all complex numbers. What we learned then tells us now that 1 is a critical point for f. Similar considerations applied to -1 reveal that it behaves critically as well, giving us two superattracting fixed points. Moreover, deeper theory implies that these are the only critical points. Appealing to the Critical Theorem, the next order of business is to work out which points belong to their basins.

Since $f(z)$ and $F(w)$ amount to different ways of saying the same thing, the two maps are dynamically equivalent, meaning that the behavior of an orbit of z_0 matches that of its associated point w_0 and vice versa. In other words, changing coordinates converts a dynamical description in one coordinate to an account in terms of the other. A manifestation of this equivalence between maps is that 0 and ∞ are superattracting fixed points relative to F. The point of changing coordinates is to ease the burden of identifying F's basins (at the diagram's bottom right). Indeed, we more-or-less already found these sets in an earlier example. Watching what happens to the lengths of orbit points for an initial condition w_0, you can readily see that the orbit approaches 0 or ∞ if the length of w_0 is less or greater than one. Summarizing, the basin of 0 or ∞ is the interior or exterior of the unit circle. For points that reside *on* the unit circle, their orbits remain there, behaving in complicated, chaotic ways.

In the case of F, the circle is the map's *Julia set*, which you can think of as the collection of points where the dynamics is wild. Our interest, however, lies with those points whose orbits are particularly stable—generally known as the *Fatou set*. An element in a basin of attraction is one species of Fatou point. All of the maps that contribute to equation-solving have a key property: every point in a map's domain belongs to either a superattracting basin or the Julia set. Moreover, the dynamics is reliable, meaning that the basins are massive.

In our diagrammatic journey's ultimate stage, we uncover the basins for f in z-space. All it takes is to push the basin structure that we found in w-space through the transformation from w to z (leftward arrow at bottom). This process undoes the switch from z to w so that the unit circle in w-space becomes the vertical line in z-space. As the coloring shows, the circle's interior and exterior are "unwrapped" onto the right and left halves of the real plane. Finally, conclude that orbits beginning in the left half-plane approach -1 and those whose starting point belongs to the right half-plane tend to 1. The vertical line (in the real sense) that separates the half-planes inherits chaotic dynamics from the unit circle in w-space.

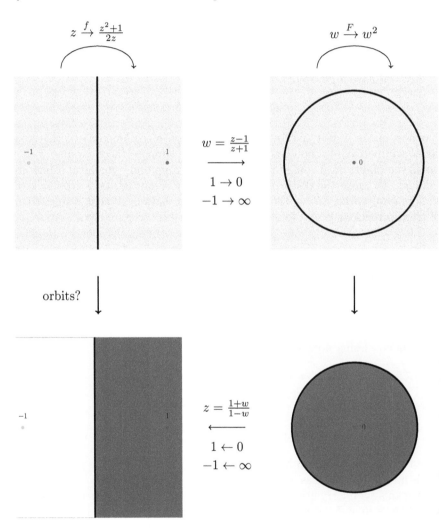

Figure 5.13 Understanding global dynamics. The goal is to start from the top left state and reach the bottom left. As explained in the text, you can get there by routing through the spaces at right using a coordinate change.

5.4   DYNAMICAL GEOMETRY AND SYMMETRY

Having the algebraic, geometric, and dynamical pieces in place, we can take the penultimate step toward construction of devices that uncover roots of polynomial equations. Unsurprisingly, symmetry is the thread that ties the three components together. In algebra, symmetry's role appears in root functions—expressions that remain the same after shuffling variables. Further instances of invariance like this are fundamental, as will soon be evident. Group actions link geometry to algebra by characterizing and quantifying symmetries associated with polyhedra and polytopes. We can now articulate and explore what it means for a dynamical system to be symmetrical.

Say you have a group of transformations

$$G = \{T_1, \ \ldots \ , T_n\}$$

that acts on an object X. For us, this collection often amounts to the rotational symmetries of a polyhedron. In addition, suppose there's a map f that moves the points on X. As indicated already, group and dynamical orbits are different. To make the distinction somewhat more formal, take a point x on X. The group orbit of x is the collection of elements in X that results when all transformations

$$T_1, \ \ldots \ , T_n$$

in G act on x. Designate this subset of X by

$$GO(x) = \{T_1(x), \ \ldots \ , T_n(x)\}$$

where $T_1(x)$ is the point to which T_1 sends x, and so on. Due to group properties, such a set remains the same if you generate the group orbit for any of its members. Using notation,

$$GO(x) = GO(T_1(x)) = \ \ldots \ = GO(T_n(x)).$$

To restate, the dynamical orbit of x is an ordered set

$$DO(x) = (x, f(x), f^2(x), \ \ldots \ ).$$

Note that when the group has a finite number of elements, the group orbit's size is also finite, whereas a dynamical orbit is always infinitely long.

A map's symmetry arises from its association with a group action, the idea being that a map f moves points in a way that respects the action performed by a group G. As far as an action is concerned, points in a group orbit are equivalent to one another; it's as though the transformations in G treat the orbit as a single entity. Indeed, you can think of each $GO(x)$ as one point in a separate *orbit space*. So, what happens when you feed a group orbit into a symmetrical map, a process that can be expressed as

$$f(GO(x)) = \{f(T_1(x)), \ \ldots \ , f(T_n(x))\}?$$

The answer, portrayed in Figure 5.14, is that you get a group orbit, namely,

$$f(\mathrm{GO}(x)) = \mathrm{GO}(f(x)) = \{T_1(f(x)), \ \cdots \ , T_n(f(x))\}.$$

Put tersely, this **symmetry condition for maps** declares that group orbits map to group orbits. Knowing where such a map sends one point in an orbit automatically determines where the remaining points go.

$$f(\mathrm{GO}(z_1)) = f(\mathrm{GO}(z_2)) = f(\mathrm{GO}(z_3))$$

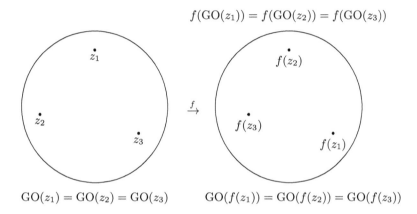

$$\mathrm{GO}(z_1) = \mathrm{GO}(z_2) = \mathrm{GO}(z_3) \qquad \mathrm{GO}(f(z_1)) = \mathrm{GO}(f(z_2)) = \mathrm{GO}(f(z_3))$$

Figure 5.14 What is symmetry—for a map? The left ring depicts a group orbit $\mathrm{GO}(z_1)$ of a point z_1 and encircled on the right are the points where a symmetrical map f sends the orbit. Symmetry in a map means that the set on the right is also the group orbit of $f(z_1)$.

The issue at hand concerns the construction of maps with properties that are suitable both geometrically and dynamically. Before scrutinizing the topic, we should pause to wonder why symmetrical maps are appropriate tools for the purpose. Recalling the Galois Doctrine, an equation-solving process must break a polynomial's symmetry. We'll see how a symmetrical map's dynamics provides a way to take this essential step. The lesson here is one learned earlier, when labeling an equilateral triangle: symmetry-breaking is best accomplished with an object that exemplifies the symmetries in question. There's also the matter of our method's elegance, which is for the reader to decide.

With that background, the time has come to engineer maps, classified by symmetry group, that drive algorithms designed to obtain roots of polynomials. Guiding us will be the configurations treated in Section 5.2. Most cases include a formula for the respective map, but technical discussion overall is modest. The goal is to acquire a sense of how the various pieces fit together. Knowing the explicit form that a map takes is not essential to understanding its geometric attributes and dynamical behavior. Nonetheless, you might find a pleasing aesthetic in the algebraic descriptions—along with the geometry and dynamics.

S_2

Solving a quadratic equation requires a map that reliably "finds" two points on the complex line. We can accomplish the job with a procedure dating back at least to Isaac Newton. To solve the equation

$$P(z) = z^2 + bz + c = 0,$$

Newton's method iterates the map

$$n(z) = \frac{z^2 - c}{2z + b}.$$

Figure 5.15 shows how Newton's map arises geometrically.

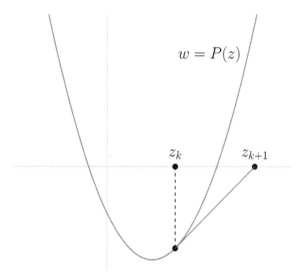

Figure 5.15 Constructing Newton's method. Think of the horizontal z-axis as the "dynamical space," where the iteration takes place and take z_k to be some point in an orbit. At the location $(z_k, P(z_k))$ on the graph of $w = P(z)$, attach the (red) line tangent to the curve. Obtain the next orbit point z_{k+1} where the line crosses the z-axis. When the z-axis is a real number line, the diagram can be read literally, whereas it has a figurative interpretation in case the axis is a complex line.

For a degree-2 polynomial, $n(z)$ has reliable dynamics, due to the roots of P being both fixed and critical. Moreover, since there are no other critical points, the orbit of a random starting point homes in on one of P's roots, as the Critical Theorem demands.

Things are simple enough to attain an exact description of the basins for these superattracting points. In the spirit of algebra, pretend that you know the roots of P; call them r_1 and r_2.[3] They lie on the complex line and, so long

---

[3]We used z_1 and z_2 in dynamical orbits.

as they're not equal, can be connected by a line segment. Such a plot appears in Figure 5.16. Following the same tactic as in Figure 5.13, use a transformation of coordinates that sends r_1 to 0 and r_2 to ∞. As in the earlier example, the map takes on a simple form in the new coordinate w:

$$N(w) = w^2.$$

The line perpendicular to and bisecting the segment between the roots transforms into the unit circle. Dynamically this is the same situation at which we arrived in the prior example: 0 and ∞ are superattracting fixed points whose respective basins are the interior and exterior of the circle. Transforming back to the original space, each basin becomes the real half-plane of the same color. Because the dynamics of n and N are equivalent, the basins of r_1 and r_2 are the half-planes while chaos reigns on the boundary line that separates them. As for the noted similarity between Newton's method in general and the specific map f that served in our example of changing coordinates, f happens to be what results when you apply Newton's method to the polynomial

$$z^2 - 1.$$

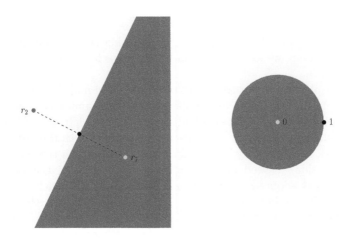

Figure 5.16 Solving the quadratic with the global dynamical behavior of Newton's method. The z-space (left) transforms to w-space (right) where the global dynamics is easily worked out and then transformed back to the original space.

S_3

Associated with the group that shuffles three objects is a double triangular pyramid—a beach ball polyhedron on display in Figure 5.5. With respect to its rotational symmetry group, each of the structure's six triangular faces are equivalent. That is, as you apply the rotations to any one triangle, that face moves onto all of the others. Since we're interested in a map that respects the beach ball's symmetries, we need to consider only how it acts on one face.

Using tools drawing on algebra and geometry, we arrive at a map that extends the method of Newton and bears the name Halley, a contemporary. Situating the S_3 polyhedron as in the figure amounts to selecting a coordinate system with which we can interpret its features. The form taken by the map in *Halley's method* for the general cubic polynomial is not very illuminating; let's refer to it as $h(z)$. Note that each cubic gives rise to a specific Halley map.

By changing coordinates, we can glean how such a map behaves by scrutinizing one especially nice case, namely, when the equation is

$$w^3 + 1 = 0.$$

Now, you can solve this equation straightaway—no dynamics is called for. However, such is not the case for most cubics. Fortunately, the Halley map for a randomly-selected cubic is dynamically equivalent to the degree-4 Halley map $H(w)$ corresponding to the special polynomial $w^3 + 1$:

$$H(w) = \frac{w(w^3 - 2)}{2w^3 - 1}.$$

The power in this relationship is that, by grasping the dynamical properties of one special map, we can understand Halley maps for nearly all third-degree polynomials.

A core aspiration in our treatment of symmetrical maps is the discovery of elegant geometric properties exemplified by their dynamics. So now, consider one triangular face F of the beach ball as shown on the left side of Figure 5.17 as a green-shaded triangle in the southern hemisphere with the visible critical points in red. Feeding every element in F into H results in the shaded region on the right side—viewed in three different ways. At top, the viewpoint is identical to that adopted at left. The middle looks at the beach ball from the other side while the bottom viewpoint is directly above the north pole. As for the operation's net effect, the single face F maps over four triangular faces: F itself (top) and the three faces that form the northern hemisphere (bottom). The view at middle right shows that nothing maps to the other two southern faces. That one triangle stretches to cover four is an emblem of the map's degree.

The scenario here is similar to what occurs with quadratics in that each root of a randomly-selected cubic is fixed and superattracting under its associated Halley map. One difference is that the cubic's roots count double as critical points—a condition with geometric consequences, as we will see shortly. Since the three roots are the only critical points, we take from the Critical Theorem that the orbit of an initial condition chosen at random belongs to one of three superattracting basins. As far as a map's dynamics is concerned, the way it behaves at critical points is, well, critical. On the beach ball, four edges come together at a critical point to form four right angles, one of which is exemplified by the shaded triangle at left in the figure. When the map H acts on a piece of the shaded triangle near the vertex—which is fixed, that right angle opens up to an angle three times as big. This signature of a critical point can be clearly seen in the top and middle views on the figure's right side where the shading covers three right angles. Because of its symmetry, H does the same thing to each of the critical points. Tripling the angle is a geometric condition that corresponds to the algebraic fact that each critical point counts twice.

It now makes sense to inquire into the global dynamics of the special Halley map. After seeing the result for two critical points, you might guess that the basins of superattraction form three equal sectors created by three equally-spaced lines of longitude, one crescent for each critical point. Actually, a bit of theory in complex dynamics rules out such a simple outcome. The considerably more intricate state of affairs appears in Figure 5.18. These images and others to follow later are portraits of attracting dynamics that reveal computational results derived from running routines for plotting H's basins of attraction. Their underlying graphical procedure divides the respective spaces—either a complex sphere or line—into grids consisting of fine-grained cells. Systematically, they select a point p in a cell—the center, say—and then compute p's dynamical orbit. If the sequence of orbit points gets attracted to one of the fixed critical points, every cell associated with members of the orbit acquires a color that's assigned to the superattracting point. This technique identifies points that belong to a basin, at least approximately, and gives more precise results with a finer grid.

The plots assign cyan, magenta and yellow to the three critical points. (Note that the right image shows the critical locations as red, blue, and green for purposes of comparison to the prior Figure 5.5.) When the routines run over all cells, there's no point that fails to reside in one of the three known basins, empirically confirming an outcome that the Critical Theorem requires. Amusingly, the complex line plot at right can be taken as a nearly literal dartboard for which a random throw lands on one of the three colors. For the map h associated with the original cubic, the graphical rendering of its basins and corresponding dynamical classification *qualitatively* agree with the description given for the singular map H. If you plot the basins for a typical sample h, their form is obtained by systematically stretching and shrinking H's basins.

Fractal behavior is also evident in the graphics here. Take, for instance, the large yellow "wedge" which contains the critical point c noted as red in the right image. This piece is the immediate basin of c and, as such, maps onto itself while each of its points has an orbit that homes in on c. All of the other infinitely many yellow regions eventually maps on top of the wedge to which c belongs. What's more, the central point (colored orange)—which happens to be 0 for the coordinates in use—is fixed and repelling. As H

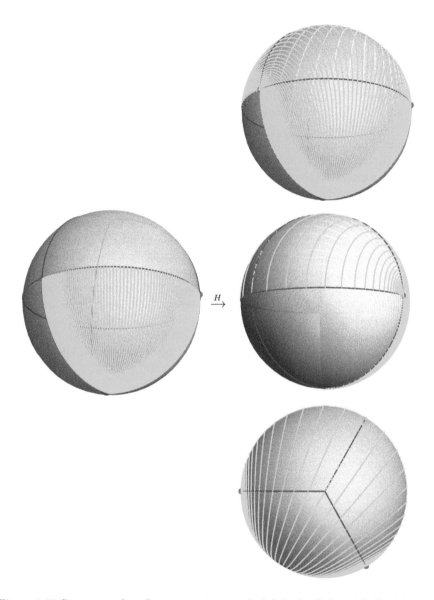

Figure 5.17 Geometry of an S_3-symmetric map. At left is the S_3 beach ball with one triangular face shaded with non-intersecting segments. The shading in plots on the right reveal what happens to points in the shaded face when the Halley map H is applied. There are three different views of one sphere.

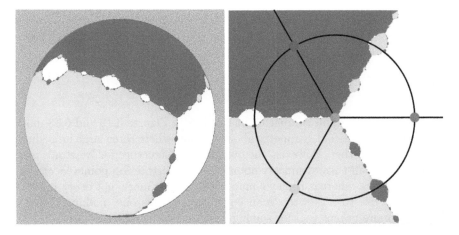

Figure 5.18 Global dynamics on a double triangular pyramid. Basins for three roots of a cubic under the Halley map H on the complex sphere and line. (The software package *Julia*[37] produced the graphics at left and the basin plot to the right resulted from running [24].)

undergoes iteration an infinite supply of yellow regions emanates from 0 and, with each step, enlarge and move away from 0. The same phenomenon occurs for the cyan and magenta basins. A repelling fixed point is also called a *source*, terminology for which this map supplies a rationale.

What can be said about the points lying outside the three basins? You come across such locations if you travel from one basin to another—cyan to magenta, say. The totality of these points—known as the map's Julia set— forms a kind of "dust" at the basins' boundaries. Its powdery nature differs sharply from the circular set of Julia points for Newton's map applied to quadratics. Once again, there are an infinite number of Julia points, each of which maps to another in such a way that chaotic dynamics ensues. Because the Julia dust is scattered so thin that it contains no part that has positive density, there's no chance of selecting a dust particle at random.

Next, we note the symmetry that emerges with the basins. Viewing the spherical model, the six rotational symmetries of the double triangular pyramid are present, provided that we *ignore* the basins' colors; that is, if we give all basins the same color. Would such a circumstance mean that the entire sphere is monochromatic, with a massive gain in symmetry? No, it would not, due to the existence of Julia points that are not residents of any basin. The Julia set is a collection that forms infinitely thin clusters, rendering it effectively invisible. If we acknowledge the basin colors, the symmetry breaks. With this observation, we come across an example of a price that can be paid in order for symmetry to exist: loss of color vision might lead to an increase in symmetry. Briefly, see less color, see more symmetry.

For an instructive comparison regarding the significance of a map's symmetry, let's consider Newton's method for $z^3 + 1$. Using the same coordinate as in the Halley case, we get a dynamical system of degree three:

$$z \to \frac{2z^3 - 1}{3z^2}.$$

Disregarding colors and comparing basin plots in Figures 5.19 and 5.18, both images reveal three-fold symmetry about the pole where three large immediate basins come together. However, the overall basin structure for Newton's map fails to enjoy half-turn symmetry about the superattracting points located in the large pieces, whereas Halley's map does. What's more, not every random cubic gives rise to a Newton map equivalent to this one, making Halley's method a more potent tool for solving degree-three equations.

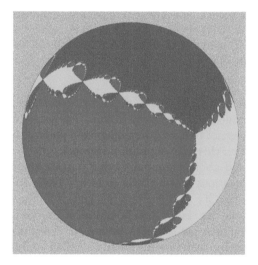

Figure 5.19 Global dynamics on a double triangular pyramid whose north and south poles are distinguishable. Basins of three roots for Newton's method. (Image is the product of [37].)

S_4

Looking at the spherical configurations of the cube or octahedron (Figure 5.6), recall that their rotational symmetries completely shuffle four pairs of antipodal vertices or face-centers. If we situate the cube in three real dimensions with two of its six face-centers on each of the three coordinate axes and then stereographically project the cube, we get six projected face-centers on the complex line:

$$\{\infty, 0, -1, 1, -i, i\}.$$

The four pairs of antipodal vertices project to four sets of two complex numbers:

$$\left\{\frac{(1+i)\sqrt{3}}{3-\sqrt{3}}, -\frac{(1+i)\sqrt{3}}{3+\sqrt{3}}\right\} \qquad \left\{\frac{(1-i)\sqrt{3}}{3+\sqrt{3}}, \frac{(1-i)\sqrt{3}}{-3+\sqrt{3}}\right\}$$

$$\left\{\frac{(1+i)\sqrt{3}}{-3+\sqrt{3}}, \frac{(1+i)\sqrt{3}}{3+\sqrt{3}}\right\} \qquad \left\{\frac{(1-i)\sqrt{3}}{3-\sqrt{3}}, -\frac{(1-i)\sqrt{3}}{3+\sqrt{3}}\right\}.$$

Among the cube's rotational symmetries, each of these pairs has a symmetry group equivalent to S_3 accompanied by a degree-six function that remains the same when acted on by members of that group. Designate these S_3-invariant polynomials Q_1, Q_2, Q_3, and Q_4. The full cube group shuffles them in the same way that S_4 rearranges four objects. For instance, say that M is a Mobius transformation that corresponds to one of the cube's quarter-turn symmetries. When M "acts on" them, the Q functions run through a cycle whose length is four:

$$Q_1(M(z)) = Q_4(z) \qquad Q_4(M(z)) = Q_3(z)$$
$$Q_3(M(z)) = Q_2(z) \qquad Q_2(M(z)) = Q_1(z).$$

Next, we derive a basic tool, namely, algebraic expressions that don't vary when the S_4-equivalent group of the cube's rotational symmetries acts on them. Take the polynomial $F(z)$ that returns 0 when evaluated at each face-center. That is,

$$F(0) = F(-1) = F(1) = F(-i) = F(i) = 0$$

and the form that results is

$$F(z) = z(z^4 - 1).$$

Arranging for F to "vanish" when $z = \infty$ requires a more general system of two coordinates, one that the cited research papers use. Following the same course with the vertices produces a degree-eight polynomial

$$G(z) = z^8 + 14z^4 + 1.$$

What makes these expressions invariant under the full group of cube symmetries is that they each correspond to a complete group orbit, meaning that the vertices and face-centers shuffle among themselves.

Bringing algebraic machinery to bear on these "basic invariants" yields two rational maps with 24-fold symmetry whose degrees are one less than that of the respective polynomial. From $F(z)$, whose "true" degree is six due to the role played by ∞, we get a degree-five map:

$$f(z) = -\frac{z(z^4 - 5)}{5z^4 - 1}.$$

Applied to $G(z)$, the calculation gives

$$g(z) = -\frac{7z^4 + 1}{z^7 + 7z^3}.$$

Although their specific form is not very illuminating, the reader might find a source of aesthetic delight in the simplicity enjoyed by these formulas as well as by those to follow.

Along with their algebraic elegance, these symmetrical maps exhibit beautiful geometric behavior. For the "cube-map" f, a square face A of the cube stretches over itself and its four adjacent faces, covering everything but the face antipodal to A. One face mapping over five signifies the map's degree. The process also includes some twisting so that each vertex of A maps to its antipode. At a vertex, two edges of a face form an angle of 120°; remember, we're working on the sphere. When f is applied, this angle opens to one that takes up two 120° angles. (A physical model of a cube can help with visualizing what happens.) This "wrapping geometry" indicates that the eight vertices are critical points. Since a degree-five map is critical at exactly eight locations, the vertices make up the entire set of critical points. Hence, the Critical Theorem is in force, implying that antipodal pairs of vertices are superattracting cycles whose basins fill the sphere in the probabilistic sense discussed previously. This reality is captured in Figure 5.20 by the plot at left.

Running through a similar discussion for the octahedral-map $g(z)$, each triangular face B maps over seven—again, the map's degree, excluding only the face antipodal to B. Here, too, vertices have period two as the map sends each one to its antipode. Furthermore, these points alone are critical, which puts the Critical Theorem in play. Since a 90° angle between two intersecting edges gets mapped to an angle equal to 270° = 3 · 90°, each vertex counts as critical twice. At an infinitely small scale, the map resembles a third-power operation as it shrink-wraps a tiny region around a vertex. At right in Figure 5.20, the basin plot reveals the Critical Theorem's consequences.

A_5

For this group, the discussion runs in parallel to that of S_4, except that we now work with the exquisite 60-fold symmetry enjoyed by the dodecahedron and icosahedron. Denote this collection of rotational symmetries as G_{60}. A key feature of the icosahedral structure is the decomposition of its twenty face-centers or dodecahedral vertices into five classes of four points. Each quartet is associated with 1) a subgroup of G_{60} that moves the four elements as vertices of a regular tetrahedron and 2) a degree-eight function that doesn't change when this subgroup is applied. To make clear the parallels to structural properties associated with other groups, call these tetrahedral invariants Q_1 to Q_5. Both the tetrahedra and their companion functions experience even shuffling when G_{60} acts on them.

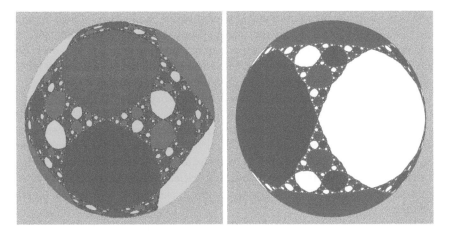

Figure 5.20 Global dynamics with cubic and octahedral symmetry. At left is the basin plot for the cube map. Four colors indicate that the plotting routine detects that many basins. Eight large immediate basins appear, each of which holds one vertex in a superattracting cycle of length two—note the color match between two green and red regions on opposite pieces of the sphere. For the degree-seven map characterized on an octahedron, the plot on the right shows three basin colors associated with cycles consisting of antipodal vertices. Note the magenta immediate basins at the north and south poles. In both plots, complex fractal structure appears within the regions outside of the immediate basins. (Images are the product of [37].)

As for polynomials that are invariant under the action of the full icosahedral group G_{60}, you can apply a method like the one used in S_4 algebra that produces two basic expressions connected to the special group orbits of vertices and face-centers. 'Basic' means that application of elementary algebraic procedures to these two polynomials generates all such invariants. Reusing notation, their elegant forms have degrees eleven and twenty respectively:

$$F(z) = z(z^{10} - 11z^5 - 1)$$
$$G(z) = z^{20} + 228z^{15} + 494z^{10} - 228z^5 + 1.$$

As before, ∞ is a vertex so that F's degree in the single coordinate z is eleven rather than twelve, the number of vertices.

Utilizing the same technique as in the S_4 case gives maps with icosahedral symmetry as well as pleasing numerological forms:

$$f(z) = -\frac{z(z^{10} - 66z^5 - 11)}{11z^{10} - 66z^5 - 1}$$

$$g(z) = -\frac{57z^{15} + 247z^{10} - 171z^5 + 1}{z^{19} + 171z^{14} + 247z^9 - 57z^4}.$$

These maps naturally associate with the dodecahedron and icosahedron respectively. They exhibit geometric behavior analogous to what takes place in the cube and octahedron maps.

When f acts on the dodecahedral configuration, a pentagonal face A maps over the eleven faces—hence, the degree—outside of A's antipodal face. At each vertex, the $120°$ angle on a pentagon's interior expands to cover two such angles at the antipodal vertex. Since the periodic vertices account for all twenty critical points, superattracting dynamics dominates globally.

As for the degree-19 map g, a now familiar story unfolds. Each triangular face stretches over nineteen, namely, all faces except the one that's antipodal to the initially selected face. Since five triangles surrounding a vertex cover $360°$, two edges of a triangle form a $72° = 360°/5$ angle on the sphere. When g is applied, one $72°$ wedge at a vertex shrink-wraps onto four such wedges attached to the antipodal vertex. Counting critical points, each vertex is worth four and, under infinitely strong magnification, the map performs like fourth-powering.

Once more, by the Critical Theorem and as Figure 5.21 unveils, a random dynamical orbit for f or g will find one of the superattracting cycles.

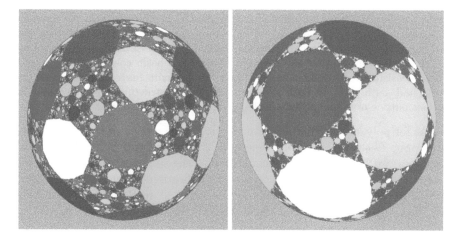

Figure 5.21 Global dynamics with dodecahedral and icosahedral symmetry. However many superattracting basins appear, there are an equal number of colors. For the dodecahedron map (left) and icosahedron map (right), there are ten and six, with immediate basins structured in antipodal pairs. As for the powdery fractals of non-basin points that constitute the maps' Julia sets, their intricacies are consigned to imagination. (Images are the product of [37].)

A_6

We now relocate to a complex space with two dimensions so that it takes that many variables (z, w) to describe geometric features. Each system of six conics is defined by degree-two equations; let them be given by six conditions:

$$Q_1(z, w) = 0 \quad \text{to} \quad Q_6(z, w) = 0.$$

Note that the degree of a two-variable polynomial uses the sum of exponents for each term. Another set of six degree-two expressions $R_1(z, w)$ to $R_6(z, w)$ defines the other system of conics. In order to cook up six functions that Valentiner's group shuffles in the manner of A_6, you need to cube the Q or R functions. In the usual way, summing either set of these degree-six polynomials—second-degree raised to the third power—creates a Valentiner invariant with a nice form in coordinates that strike a balance between the two systems of conics:

$$
\begin{aligned}
F(z, w) &= Q_1(z, w)^3 + \ldots + Q_6(z, w)^3 \\
&= R_1(z, w)^3 + \ldots + R_6(z, w)^3 \\
&= 10z^3w^3 + 9z^5 + 9w^5 - 45z^2w^2 - 135zw + 27.
\end{aligned}
$$

Classic theoretical techniques can derive from F two additional invariants G and H that complete a set of basic (generating) polynomials in the sense used previously; their respective degrees are 12 and 30. Due to length, their expressions are withheld.

Constructing maps with Valentiner symmetry takes a bit of finesse on top of theory that's more advanced than in the one-variable setting. Discovering and proving the properties that a higher-dimensional dynamical system seems to possess on experimental grounds is also considerably more challenging.

Two discoveries that meet our aesthetic criteria stand out. Pertaining to the first, a family of maps occurs in the intriguing degree of nineteen. What this fact makes possible is for each icosahedral conic in both systems to map over itself in exactly the same way as the icosahedron map pictured at right in Figure 5.21. This hunch was confirmed by explicit computation [7]. As noted in the prior discussion of Valentiner geometry and sketched in Figure 5.8, the superattracting vertices are the sites where two conics in different systems meet. Each conic in one system possesses twelve points of this type, giving a Valentiner orbit of 72 such points. Viewed dynamically, a conic attracts nearby points that lie in the direction away from the conic. Ultimately, the dynamical orbit of such an initial condition accumulates at a pair of antipodal vertices for two conics belonging to opposite systems. In this way, the twelve conics act as *attracting sets* that guide dynamical orbits to icosahedral vertices on pairs of conics. Empirical evidence strongly supports the conjecture that these superattracting 72 points have basins that fill a two-dimensional dartboard.

The second map's special property concerns how it behaves on the 45 reflection lines. Because each line performs as a mirror relative to some group element, theory requires a Valentiner-symmetric map to rearrange points that reside on the line. In addition and decisively, the critical points for the map in question coincide with the mirror lines, a condition that requires the degree to be 31. A dramatic difference between dynamics in one or several dimensions pertains to critical points: in the former, there are finitely many whereas, in the latter, an infinite number show up. In two dimensions, the notion of critical point as a shrink-wrapping location generalizes from its one-dimensional counterpart. A higher-dimensional version of the Critical Theorem applies to this rather special situation in which the entire collection of critical points maps on top of itself. Intersections between the mirrors are locations that superattract in all directions. If there were a place outside the superattracting basins away from these critical crossing points where Julia dust has some thickness, iteration would expand and spread that density across space, something that cannot occur given that basins exist. Similar considerations also apply to one-dimensional maps and lend support to the Critical Theorem. In the light of this theory, we see that the dynamics here is reliable.

Since this map runs in an environment with four real dimensions, plotting a comprehensive picture of its basins is infeasible. What we can do is narrow attention to lower-dimensional sets that map onto themselves and then graphically compute basins for the dynamical system that results from restricting to a domain in the lower dimension. Two spaces of this type exist for the degree-31 map currently under study: 1) 45 complex lines of reflective symmetry and 2) 36 spaces describable as a sphere whose pairs of antipodal points are glued to one another—a surface that you can visualize as a hemisphere whose opposite equatorial locations are sown together. Basin plots on such sets appear in Figure 5.22. At left, you see sixteen basins for the map's dynamics restricted to one of the 45 lines. A look at the right image shows eleven basins on one of the 36 hemispherical surfaces. Each immediate basin—the large zones—contains a superattracting site where some of the 45 lines cut across the pictured space. Since each collection of spaces is a group orbit—lines and hemispheres, the map's Valentiner symmetry ensures that the structure assumed by the basins is identical for each member of the respective orbit.

Juxtaposing the plots exposes a qualitative difference between the dynamical actions on distinct spaces. For a complex line, basins are clustered around points lying in a basin's boundary. In fact, every point not in a basin is surrounded by regions belonging to *all* basins [2]. Such is not the case when iteration takes place on the hemispherical surface. Another salient contrast is the mottled appearance produced by basin pieces on the hemisphere. What's going on here is a structural disparity between complex and real spaces that produces distinct dynamical behavior in the two settigns. The sphere and sown-up hemisphere are both surfaces. However, due to the presence of $i = \sqrt{-1}$, the former has a notion of complex multiplication that the latter is wanting.

5.5   SOLVING EQUATIONS BY ITERATION

Obtaining the roots of a degree-d polynomial ultimately comes down to a symmetry-breaking choice among

$$d! = d \cdot (d-1) \cdot \ldots \cdot 3 \cdot 2 \qquad \text{or} \qquad \frac{d!}{2}$$

indistinguishable objects realized as ordered lists—that is, point-coordinates—of d complex numbers. The particular number of points depends on whether

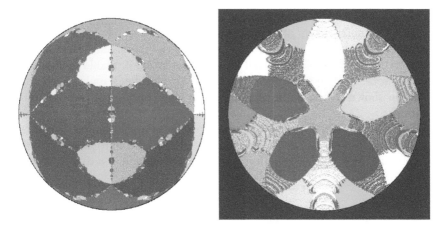

Figure 5.22 Global dynamics on two spaces preserved by a degree-31 map with Valentiner symmetry. To the left is a rendering of superattracting basins on a complex mirror with four-fold rotational symmetry. The right side shows basins on the ten-fold symmetric hemispherical space described in the text. Matching colors around the bounding circle reveals the way opposite points are joined. (Images are the product of [37] (left) and [24].)

we use the complete or even shuffle group. Not being able to tell these objects apart is a consequence of the action performed by groups that shuffle the coordinate entries. We then find equivalent groups acting on low-dimensional spaces by rearranging pieces of various geometric structures. Finally, on top of this geometry we build dynamical systems that respect the symmetry exemplified by the relevant structural features.

From this vantage, a mechanism that breaks symmetry is now apparent. It comes into play when you randomly select an initial condition for a dynamical orbit. Appealing to the metaphor of a circular dartboard, the outcome of a random throw—which won't hit the precise center—has less symmetry than the board had originally without a dart stuck in it. But, there's a vast infinity of starting points from which to choose. We require a means of connecting a chosen point to a polynomial's roots. This crucial task is fulfilled by map iteration along with some algebraic machinery associated with representations of shuffle groups.

Quadratic

Imagine the state of affairs depicted in Figure 5.16 as a dartboard. There's a 100% chance that a random throw lands in one of the basins. (Since this is a mathematical board, you don't have to worry that it extends to infinity.) This outcome is assured despite the fact that an infinite number of points belong to the line between basins. There's no chance for a randomly-chosen initial condition to belong to that specific infinitely thin line.

As explained in Section 5.2, the non-trivial transformation in the two-element symmetry group of a quadratic's roots gives the complex line a half-turn about the midpoint of the segment joining the roots. Since the rotational symmetry exchanges basins as well as roots, the group is unable to distinguish the two sets—like the hay bales presented to Buridan's mule. Such is the situation shown in the basin plot, provided that, once again, we ignore the coloring.

The dynamically-driven process that produces a solution to a quadratic equation $P(z) = 0$ is simple to describe. For an initial condition z_0 selected at random, use Newton's method for P to generate enough of a dynamical orbit

$$DO(z_0) = (z_0, n(z_0), n^2(z_0), \ldots, n^k(z_0), \ldots)$$

so that $n^k(z_0)$ is within a small distance from a fixed point. Such an orbit point will approximate one of P's roots to high precision. What you get is a numerical solution, unlike the symbolic expression found in the quadratic formula. Motivation for the titles of this and the previous chapters is now clear.

Cubic

Select a third-degree polynomial $P(z)$ and, as with a quadratic, make a pretense of knowing the roots r_1, r_2, and r_3. Expressed formally

$$P(z) = z^3 + bz^2 + cz + d = (z - r_1)(z - r_2)(z - r_3).$$

As before, Mobius transform from z to a new coordinate w by sending r_1 to -1, r_2 to $-V_1$ and r_3 to $-V_2$ where the values

$$V_1 = \frac{-1 + i\sqrt{3}}{2} \qquad \text{and} \qquad V_2 = \frac{-1 - i\sqrt{3}}{2}$$

were introduced in Section 4.1. The cubic now takes the special form $w^3 + 1$ with its associated Halley map

$$H(w) = \frac{w(w^3 - 2)}{2w^3 - 1},$$

thereby giving an equivalence between $H(w)$ and Halley's map $h(z)$ for $P(z)$. As with Newton's method, you don't need to know the roots of $P(z)$ in order to manufacture $h(z)$. Using the theory of invariants, it can be derived directly from the P's coefficients.

Solving $P(z) = 0$ is accomplished by calculating a random dynamical orbit

$$(z_0, h(z_0), h^2(z_0), \ldots)$$

far enough so that it "finds" one of the roots. What assures this discovery is H's reliable dynamics in w-space, which $h(z)$ imitates in z-space. As $H(w)$

reliably locates the roots of $w^3 + 1$, so does $h(z)$ lead to the solutions of $P(z) = 0$. In practice, all that need be performed is a single iteration of $h(z)$ beginning with an accidental point z_0. Theoretical considerations based on coordinate change take care of the rest.

Quartic

So far, our root-finding procedures have been *direct* in that randomly-selected dynamical orbits accumulate at a root. What makes this type of solution possible is a fundamental fact about Mobius transformations: any two or three points, but no more, can be transformed to any other pair or triple. In particular, you can send the roots of a quadratic or cubic to special locations in order to produce maps whose dynamical behavior is especially nice as well as reliable. Because a Mobius transformation amounts to changing coordinates, Newton's map for a typical quadratic and Halley's map for an arbitrary cubic are dynamically equivalent to the respective special versions. When there are four roots, the transformational wherewithal to send those points to four distinctive spots is absent. Instead, we work with the four completely shuffled functions Q_1 to Q_4 linked to antipodal pairs of the cube's vertices. Using values taken in the complex line rather than the sphere gives results with a simpler appearance.

The following discussion sketches how to build a quartic-solving algorithm around the cube map $f(z)$. In order to convey a taste of the development as well as its aesthetic qualities, a few technical details are given. A full treatment is fairly advanced and beyond the scope of this work. When the degree is five or six, articles in the research literature explain relevant theoretical and empirical aspects [13, 7].

To begin, we need a collection of quartics to solve and to that end, we can appropriate S_3 and S_4 invariants mentioned in prior discourse to design four rational functions as follows:

$$r_1(z) = \frac{F(z)^4 Q_1(z)}{G(z)^4} \qquad r_2(z) = \frac{F(z)^4 Q_2(z)}{G(z)^4}$$

$$r_3(z) = \frac{F(z)^4 Q_3(z)}{G(z)^4} \qquad r_4(z) = \frac{F(z)^4 Q_4(z)}{G(z)^4}.$$

For the sake of algebraic considerations, the numerators and denominators each have degree 32. Recalling a term used when discussing Tschirnhaus transformations, turn each of these functions into roots of a fourth-degree "resolvent" defined by

$$R_z(u) = (u - r_1(z))(u - r_2(z))(u - r_3(z))(u - r_4(z))$$
$$= u^4 + k_3(z)u^3 + k_2(z)u^2 + k_1(z)u + k_0.$$

In this setting, things are more elaborate in that two variables u and z show up. Actually, we should think of z as a *parameter* that tells you which polynomial you're talking about, while u is a standard place-holder. Notice that this factorization looks like the description in Section 1.3. The difference here is that we view the roots themselves as functions of a variable z. Multiplying the product out gives coefficients k_0 to k_3 attached to the powers of u that are root functions of r_1 to r_4. As we know, these functions don't change when the roots are shuffled. Since the cube's symmetries shuffle the roots, these rotations leave the coefficients invariant as well. Recall what this invariance means; if T is a symmetry of the cube,

$$k_3(T(z)) = k_3(z)$$

and so on. A statement related to the Shuffle Theorem implies that each coefficient can be written in terms of the basic S_4 invariants $F(z)$ and $G(z)$. After working out the relevant forms and then making the substitution

$$V = \frac{F^4}{G^3} \qquad \text{(numerator and denominator have degree 24)},$$

we arrive at a "one-parameter" family of quartics

$$R_V(u) = u^4 - 4Vu^3 + 6V^2u^2 + 4V^3(256V - 1)u + V^4.$$

Assigning a value to the parameter V amounts to selecting a quartic. As an example, set $V = 1$ so that

$$R_1(u) = u^4 - 4u^3 + 6u^2 + 1020u + 1.$$

The quartic-solving procedure about to be assembled is specifically tuned to this family of resolvents. But, you might observe, this bunch of polynomials is not exhaustive. The full set of quartics involves four parameters, not one. In a feat of algebra, it turns out that, by applying Tschirnhaus transformations, the number of parameters can be reduced from four to one. (See [16, Ch. XII].)

At this point we have quartics to solve and a cube map $f(z)$ whose dynamics reliably locates one of the cube's four antipodal pairs of vertices. What's lacking is a means of connecting the two regimes. Such a link can be forged through the parameter V, whereby we create a family of dynamical systems $f_V(w)$ each member of which corresponds to a value assigned to V and is equivalent to the special map $f(z)$. Note the introduction of a new variable w with an associated space—a development that we examine presently. You can view this process as *self-parametrizing* a collection of symmetry groups for a distorted cube in w-space. In other words, we use the cube-group consisting of rotations in z-space to create a parametric description of transformation groups each of which is equivalent to S_4. A significant and pleasing feature of this construction is that we need not compute the parametrized groups explicitly. Applying the theory of invariants, we can create the family of maps $f_V(w)$ directly. For each value of V varying over the complex numbers, you get a symmetry group for a distorted cube and a w-space on which it acts. Each space for which w is a coordinate contains a tweaked version of a standard cube whose symmetries reside in the distorted group. Every w-space realizes a cube-like configuration of eight vertices and twelve edges that form six faces. Such structures are skewed versions of a standard cube whose symmetry group consists only of spherical rotations. The distorted cubes have 24 symmetries expressed as Mobius transformations, rather than as rotations. Similar observations apply to the structure of f_V's basins, which is a distortion of the cube-map's basin plot shown earlier in Figure 5.6.

An arbitrary value for V presents us with an associated quartic $R_V(u)$ and map $f_V(w)$. Using w as a coordinate on the space where $f_V(w)$ runs, choosing a generic initial condition w_0 breaks the quartic's S_4 symmetry that's realized in the particular w-space. The dynamical orbit of w_0 then asymptotically detects one of the distorted cube's four vertex-pairs (v_1, v_2):

$$f_V^k(w_0) \xrightarrow{k \to \infty} (v_1, v_2).$$

In the skewed environment, such a pair might not be truly antipodal, but the pairing of vertices is maintained.

Having in hand a vertex-pair for the given distorted cube, our final step harvests a root of the resolvent $R_V(u)$. Employing the resources of invariant theory again, we work out a function $K_V(w)$ that returns one of R_V's roots when evaluated at a number approximating either v_1 or v_2. That is,

$$R_V(K_V(v_1)) \approx R_V(K_V(v_2)) \approx 0.$$

Taking in the algorithm's sweep, it's both robust and reliable—you can randomly select both a quartic (robustness) and dynamical orbit (reliability). As a final remark, let's take stock of what this procedure does. We begin with a polynomial whose roots $r_1(z)$ to $r_4(z)$ depend on z and then derive an equivalent form expressed in terms of a special parameter V. The problem that's actually taken on is to solve $R_{V_0} = 0$ for some specific value V_0 of V. Remember that V is itself a function of z. If you had a way to determine a value z_0 for z that produces the quantity assigned to V—that is, $V_0 = V(z_0)$, you could immediately compute the roots as $r_1(z_0)$ to $r_4(z_0)$. Coming up with such a z_0 amounts to solving an inverse problem. Mentioned early on, the difficulty stems from the condition that one value of V corresponds to many values of z. In general terms, a polynomial's symmetry is embedded in its associated inverse problem and the dynamics of an affiliated symmetrical map can be harnessed to the task of breaking that symmetry.

The method that approximates a quartic's roots is the model for equation-solving when the degree is greater than three. For clarity's sake, Figure 5.23 diagrams the algorithm's flow. Treatments of polynomials in the fifth and sixth degrees, to which we turn next, parallel the path followed in degree four. As such, the discussion presents fewer details.

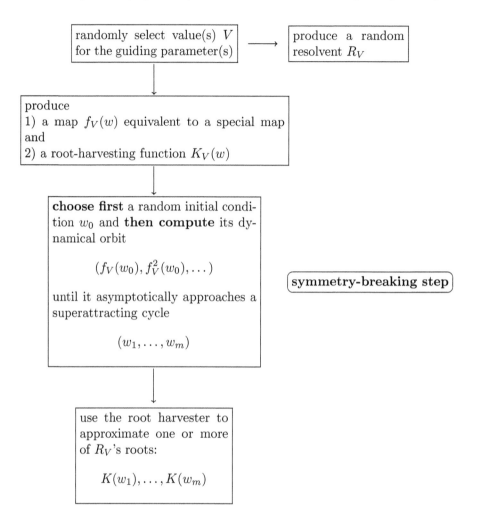

Figure 5.23 An iterative algorithm for solving polynomial equations.

Quintic

We've arrived at the historic polynomial landmark. Galois established that arithmetic and root-taking operations are not sufficiently potent to break the fifth-degree equation's A_5 symmetry. The novel piece at the heart of this result is group theory out of which arises A_5's special structure. In opening a new passage to solving the quintic, Klein connected geometric properties of the icosahedron with shuffling performed by A_5. Following Klein's lead, Peter Doyle and Curtis McMullen found a new route to the quintic's solution by exploiting icosahedral symmetry coupled to dynamics [18, 42].

In the A_5 setting (reusing some of the S_4 notation), we follow a course similar to the one taken with quartics. Declare the functions

$$r_1(z) = \frac{F(z)Q_1(z)}{G(z)} \quad r_2(z) = \frac{F(z)Q_2(z)}{G(z)} \quad r_3(z) = \frac{F(z)Q_3(z)}{G(z)}$$

$$r_4(z) = \frac{F(z)Q_4(z)}{G(z)} \quad r_5(z) = \frac{F(z)Q_5(z)}{G(z)}$$

to be roots of a fifth-degree resolvent

$$R_z(u) = (u - r_1(z))(u - r_2(z))(u - r_3(z))(u - r_4(z))(u - r_5(z)).$$

Since every coefficient of R_z relative to u is unchanged when acted on by the icosahedron's symmetry group G_{60}, the resolvent is expressible strictly in terms of the basic icosahedral invariants F and G. Defining $V = F^5/G^3$, the resolvent family becomes

$$R_V = u^5 - 40Vu^2 - 5Vu - V.$$

Our method is general since Tschirnhaus transformations can reduce the generic five-parameter quintic to this special collection in one parameter [16, Ch. XII].

In addition, V also parametrizes a set of maps $f_V(w)$ each of which is dynamically equivalent to the dodecahedron map $f(z)$. When $f_V(w)$ runs on a random initial point w_0, the orbit asymptotically reveals a pair of points—call them w_1 and w_2. On the distorted icosahedron in the space whose coordinate is w, these locations act as antipodal vertices—critical points for $f_V(w)$. They also belong to distinct clusters of four points that form two tetrahedra shuffled by the particular G_{60} that's acting on w-space.

Finally, compute two roots of $R_V(u)$ by activating the V-parametrized root-harvesting function $K_V(w)$ and inputting the iteration's output:

$$K_V(w_1) \quad \text{and} \quad K_V(w_2).$$

This map's failure to provide output from which all five roots can be harvested stems from the fact that its dynamics only *partially* breaks 60-fold symmetry. The group orbit of the superattracting twenty dodecahedral face-centers is not shuffled by G_{60} in 60 distinct ways as A_5 does to five quintic roots. Similar considerations hold for the cube map where a dynamical orbit locates the eight-point orbit created by the cube's rotational symmetry group. This fact implies that the iterative algorithm based on $f(z)$ harvests two roots. Overcoming the symmetry *completely* requires a map whose superattracting set consists of 60 periodic points (24 in the quartic setting) so that the shuffling behavior of the five roots under A_5 exactly matches that of the critical points.

This deficiency has been overcome by a more recent discovery that produced a special class of degree-31 maps each having 60 periodic critical points [11, 13]. The resulting dynamics thereby breaks A_5 symmetry fully and, in one iterative go, allows for all five roots of $R_V(u)$ to be computed. For some of these maps—naming a select one h, every critical point belongs to a cycle whose length is five. In Figure 5.24, such cycling is clearly seen in the twelve "fans" with five blades. For a given color, each blade contains one point in a superattracting cycle the elements of which belong to five distinct clusters of twelve octahedral mid-edges. Now, when you run the parametrized version $h_V(w)$ of $h(z)$, the output approximates a cycle:

$$(w_1 = h_V(w_5), w_2 = h_V(w_1), w_3 = h_V(w_2), w_4 = h_V(w_3), w_5 = h_V(w_4)).$$

The points in this cycle belong to five distinct octahedra—one for each fan-blade—each of which permits one root of R_V to be harvested:

$$K_V(w_1), \ K_V(w_2), \ K_V(w_3), \ K_V(w_4), \ K_V(w_5).$$

Note the reuse of K_V, under the assumption that it's tuned to the dynamics of h_V.

Figure 5.24 Global dynamics for a degree-31 map with icosahedral symmetry. The plot shows a disk (not a sphere) onto which a complex line has been projected. There are twelve basins (colors). Featuring prominently are the immediate basins, consisting of five petal-like regions of one color attached to a central point with five-fold symmetry. Applying $h(z)$ causes the petals to cycle like the blades of a spinning fan. (Image is the product of [24].)

Sextic

An equation-solving routine can be built around the degree-19 map that pre-serves conics. However, we'll discuss how to do so only with the map in degree

31 for which the 45 mirror lines are critical—call it $h(z)$. The collection of lines thereby forms a superattracting set that reliably guides a random dynamical orbit to a place where mirrors cross. This attraction plays out as a two-step affair in which an orbit experiences attraction *onto* a line L and then *along* L, asymptotically approaching one of sixteen critical points where the other 44 lines gather. In addition to computing a root by drawing on which point ultimately attracts an orbit, we can extract a second root by knowing which line captures an orbit [10].

Working out the details of a sextic-solving procedure tracks developments in lower degree, excepting the higher dimension and number of parameters used to reference sixth-degree resolvents, degree-31 maps, and root-harvesting functions. For Valentiner parameters, we again take ratios of polynomials whose degrees are equal:

$$V_1 = \frac{G(z_1, z_2)}{F(z_1, z_2)^2} \qquad V_2 = \frac{H(z_1, z_2)}{F(z_1, z_2)^5}.$$

Using machinery from higher-dimensional invariant theory and writing $V = (V_1, V_2)$ for short, we arrive at a family of resolvents R_V, maps $h_V(w)$, and root harvesters $K_V(W)$ and $J_V(w)$. Note the input variable for K_V is W, which we can take to be a coordinate that specifies lines in w-space. Variables $z = (z_1, z_2)$ and $w = (w_1, w_2)$ play essentially the same roles as before. As a precursor of the resolvents parametrized by V, we construct sextics with parameters z_1 and z_2. For roots, we use the six functions associated with a system of conics:

$$r_1(z) = \frac{Q_1(z)^3}{F(z)}, \quad \ldots, \quad r_6(z) = \frac{Q_6(z)^3}{F(z)}.$$

We should be aware that the nature of our fundamental inverse problem remains the same: given complex values for the parameters V_1 and V_2, find variable quantities z_1 and z_2 that satisfy the equations given above. When you have values for z corresponding to those assigned to V, you effectively possess $r_1(z)$ to $r_6(z)$. What stands in the way of determining z_1 and z_2 is the 360-fold symmetry encoded in Valentiner's group. As always, we deploy a symmetrical dynamical system that permits selection of a random point $a = (a_1, a_2)$ as a mechanism that thoroughly diminishes the group corresponding to V's prescribed value.

With V and a chosen, the dynamical orbit homes in on both a line L and a point b on L in the space described by w:

$$h_V^k(a) \xrightarrow{k \to \infty} b.$$

Given these two data points, root harvesting finishes the job, yielding two roots:

$$K_V(L) \quad \text{and} \quad J_V(b).$$

Heptic

In a development that closely parallels the one for sextic equations, we can estimate roots of seventh-degree polynomials whose symmetry group is a collection—found by Klein—of 168 two-dimensional transformations [9]. Solving the general heptic equation demands a deep understanding of the action of A_7 on three-dimensional complex space—a challenging task that's incomplete at the time of this writing.

Higher degree

The only route to an iterative solution of an equation whose degree is greater than seven is through a group action that shuffles coordinates with n entries. Recall that this is the standard representation of a full shuffle group S_n on a complex space whose dimension is $n-2$. Even though these actions are generic, they harbor maps whose dynamics is geometrically special. Transformations in the standard representation of S_n not only shuffle a certain collection of n points, but also rearrange a set of $(n^2 - n)/2$ hyperplanes—spaces whose dimension is one less than that of the entire space—that is, $n - 3$. These hyperplanes also serve as higher-dimension complex mirrors relative to certain members of the group that realizes S_n's structure. In the Valentiner representation of A_6, the 45 mirror lines and their attendant reflections are an example.

A search for interesting dynamical systems that respect standard representations as symmetry groups uncovered an infinite family of maps. For every value of n larger than two, this family contains a unique map $f(z)$ with special properties whose degree is $n + 1$ [8]. Maps with "standard symmetry" automatically fix each reflection hyperplane as a set of points. Additionally, $f(z)$ is exceptional by virtue of having *exactly* the hyperplane mirrors—counted twice—as its set of critical points.

The configuration of hyperplanes sets up a *critical cascade* in the following sense. An $(n-3)$-dimensional hyperplane H_{n-3} is a piece of $f(z)$'s set of critical points. Since the map sends H_{n-3} onto itself, we can restrict its behavior to that space. Viewing $f(z)$ along H_{n-3}, its critical points are places where the hyperplanes other than H_{n-3} cut across H_{n-3}. Such an intersection is a space that we call H_{n-4} since its dimension is $n - 4$. What matters here is that H_{n-4} is a hyperplane inside H_{n-3} and the collection of these $(n - 4)$-dimensional spaces constitutes the critical set for $f(z)$ when restricted to H_{n-3}. We then narrow our view to f along H_{n-4} to find critical points located where the hyperplanes inside H_{n-3} intersect H_{n-4}. This descent to lower-dimension hyperplanes nested in spaces of higher dimension persists until we reach a finite set of *cascade points* lying at multiple hyperplane intersections as they superattract in every direction. The dynamical effect of this recurrence is one of reliability with basins of cascade points filling the containing $(n-2)$-dimensional space in the probabilistic sense. A multi-dimensional analogue to the Critical Theorem implies that a point picked at random belongs to the basin of a cascade point [46].

To illustrate what's going on, use the representation of S_8 on complex six-dimensional space with $28 = (64 - 8)/2$ hyperplane mirrors to which we refer as H_5. (See Figure 5.25 for a graphical interpretation.) Inside H_5, the other 27 five-dimensional spaces cut through as four-dimensional sets H_4 that constitute the critical points for $f(z)$ when its domain is restricted to H_5. Repeat the process on H_4 to obtain a map whose critical points are intersections with the remaining hyperplanes whose dimension is four. Continuing in this fashion produces maps on spaces whose dimensions diminish until they reach zero—in other words, an aggregate of points.

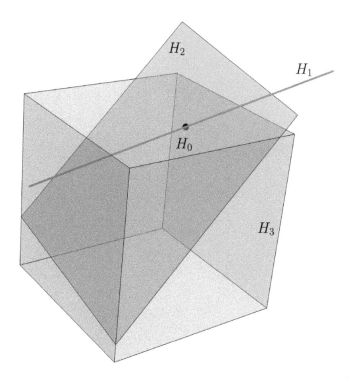

Figure 5.25 Illustrating a cascade of hyperplanes using real spaces. A three-dimensional space H_3 contains a plane H_2 that includes a line H_1 on which resides a point H_0.

Piecing together an algorithm for solving the general equation in degree n follows previous developments in the algebra of invariants. First derive a family of nth-degree resolvents R_V specified by $n - 2$ parameters that we can write as $V = (V_1, \ldots, V_{n-2})$. Next, work out parametrized maps $f_V(w)$ each of which is equivalent to the special map $f(z)$. Adopting a tactic employed when solving the sixth-degree equation, we can determine $n - 2$ roots of a chosen resolvent by converting information regarding the cascade of hyperplanes to which a dynamical orbit under $f_V(w)$ is successively attracted. These calculations require $n - 2$ root-harvesting functions, the construction of which technically exceeds the limits of this discussion.

In the S_8 example where the dimension is six, you need that many functions for harvesting roots. Compute one root by identifying the five-dimensional hyperplane H_5 that a dynamical orbit first approaches. A second root emerges from knowing which H_4 space attracts an orbit. Four more roots emerge from the subsequent cascade of spaces H_3 to H_0 onto which an orbit successively accumulates. Of course, a zero-dimensional hyperplane H_0 is a point, which is where an orbit eventually forms a cluster.

II

Beyond Equations

CHAPTER 6

Interlude: Modeling Choice

Our journey into the vast world of polynomials encountered two qualitatively different methodologies for determining roots. Any such procedure hides behind a gate bolted shut by an equation's Galois symmetry. The key to unlocking the obstruction can be found in a symmetry-breaking device. Both approaches rely, in a fundamental way, on a mechanism that opens the gate by combining a choice with a computation. They differ with respect to the order in which the two steps occur. In each case, an act of choosing itself supplies the symmetry-breaking ingredient.

This work's second part explores a variety of ostensibly non-mathematical situations that call for some kind of choice to be made. A turning of the table occurs here. In the explicitly mathematical context, choice emerged as a key feature pertaining to solving a problem. Part II treats contexts that call for a choice by drawing on insight that mathematics can provide. No claim is made that our techniques apply universally, that is, to all cases that demand a decision. Every scenario that we consider conforms to one of the models extracted from our treatment of equation-solving. Though it might not always be obvious how, the level of difficulty in deciding a question depends, to some extent, on the presence of symmetry—a condition that might lurk in the background. Faced with a difficult decision, the task is twofold:

1. recognize what symmetries are at hand

2. use one of our algorithmic models as a means of pursuing a resolution to the predicament.

Broadly, the objective is to illuminate certain aspects of decision-making in light of procedures that solve equations. Rather than build a mathematical edifice that articulates a theory for how decisions are or should be made, our account draws on mathematics as a tool with which to interpret some mechanisms that can assist with choosing. The treatment is casual by design and mostly emphasizes breadth rather than depth. Often, we take a cursory look—enough to make the point, but sometimes explore details.

CHAPTER 7

Learning to Choose

This chapter takes a tour of some choice-making devices that we liken to an algebraic style of solving equations. In such algorithms, choosing follows computation. The case studies aim to explain how appealing to a mathematical style of thinking can illuminate particular situations in which choosing is hard.

7.1 MAKING RATIONAL DECISIONS

Revisiting the impasse at which Buridan's mule finds itself, the choice is binary: go left or go right. Despite being the simplest case, a choice with but two options need not be easily made, as the mule can attest. In the context of a degree-two polynomial, there is but one relevant symmetry, namely, an exchange of the roots. You can break that two-fold symmetry with an arbitrary selection between the two indistinguishable alternatives that the quadratic formula offers.

A pure binary choice between options A and B is akin to extracting a quadratic's root when the alternatives cannot be distinguished according to the measure of some property. To the mule, an obvious way to satisfy this indistinguishability condition is to have A and B as identical stacks of hay. Less obvious is for A and B to be quantities of hay and oats such that the mule enjoys each item equally. So, regarding the property of the mule's appetite, each option has the same value. Examples of this phenomenon are easily summoned: betting red or black in roulette, fighting or fleeing when threatened by an attacker whom you can neither subdue nor outrun, eating at this or that restaurant when they're equally acceptable, voting for one of two political candidates whom you deem comparable, selling shares in a stock or not given that you have no reason to expect the price to rise or fall, and so on.

Weighing options

A special type of decision that we'll refer to as a *simple binary* occurs when B is the "negative" of A, giving options of choosing A or not. What happens if no clear preference for either A or its negative emerges? One strategy is to dig deeper, dissecting each option into pieces to which you can assign a value. Such was the method recommended by Benjamin Franklin [21]. To use his *moral algebra*, list the benefits and costs—pros and cons—either of doing A or of not doing A (denoted not-A). A better term for what Franklin had in mind might be *prudential algebra*. Only one option need be examined since a benefit or cost relative to A would have the force of a cost or benefit relative to not-A. Taking A into consideration, attach a value or weight to each item, positive for a benefit and negative for a cost. Sum the weights: preference goes to A if the sum is positive and to not-A if negative. Taking the notion of weight literally, Figure 7.1 offers a mechanical interpretation of Franklin's process. Eliminating B as an option, the diagram presents a simple binary choice.

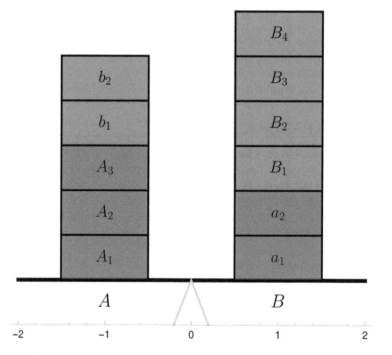

Figure 7.1 Moral algebra. Weighing a decision in the balance. Using physical weights, we can depict Franklin's method. Given a binary choice between A and B, upper-case letters represent the weight of a benefit and those in lower-case letters refer to the weight of a cost. With all weights in place, preference is given to the option towards which the balance tips.

In effect, the result of this simple calculation draws a distinction between the options, making the weight-assignment a tool that breaks the original superficial symmetry between A and not-A. But, what if adding weights turns out to yield zero or something close to zero, leaving the balance roughly horizontal? Then you cannot differentiate A from not-A relative to the property of weighting. Being able to discern a difference between A and not-A calls for some other symmetry-breaking device. A coin flip would do. But is it a rational act?

As the diagram captures, moral algebra also applies to a binary choice that's not simple. For each selection, list weighted benefits and costs.

Comparing the sums of each option's weights, prefer the one whose value is larger. Here again, stalemate can occur if the sums are equal or close to it. If there are more than two options, the procedure admits an obvious generalization.

Of course, the crux of Franklin's procedure concerns the assignment of weights. These evaluations are subjective in nature and thereby susceptible to various forms of bias—a topic to be discussed in due course. What's more, determining the weight of a benefit or cost might well entail making other choices that could be as much trouble as the one with which you began. A more general approach to this sort of problem and others is to devise a *utility function* for a range of choices and then determine which one gives maximum utility. Classical economic theory holds as axiomatic that participants in an economy are up to making such rational decisions infallibly whereas behavioral economics rejects such a hypothesis.

Partially-informed decision

What about choosing between two alternatives when a weighting process is not in the offing? For a case study, say that there are 100 lap swimmers who prefer to use a pool at a specific time. During any given session, the facility comfortably accommodates 50 users. For their preferred time, each swimmer decides whether or not to attend, hoping that when they go, no more than 49 other participants show up. Afterward, pool administrators announce how many swimmers attended a session.

This situation resembles an exercise in game theory rather than an assignment of utility to options. As such, swimmers play *strategies* that take into account their previous plays as well as the number of attendees at prior sessions. A strategy is a rule—a kind of computation that can involve randomization in some way—that dictates what play to make in a given round. Although a swimmer's choice is a simple binary one, it exhibits novelty in that interaction with other players through the agency of strategies is essential to the process.

To see how a collection of strategies can interact, we now work through a scaled-down version of the swimming pool problem. In complexity theory, strategies are categorized as *agents*, a concept that appears in our case studies on numerous occasions in a variety of settings. Say there are ten swimmers (strategies) S_1 to S_{10} and let's execute eleven rounds of play. For any one strategy S, the only data available for use are 1) "pre-plays" (previous plays) made by S and 2) prior quantities of overall attendees. Use the abbreviations 0 to indicate not-attended, 1 to stand for attended, $+$ to mean that the attendance was less than or equal to five, and $-$ to say that more than five showed up. In any given round, a player confronts one of four states:

$$(0, +), \ (0, -), \ (1, +), \ (1, -).$$

For instance, $(1, +)$ is the state of having attended in the previous round when the number of swimmers was less than or equal to five. Based on the current state, a strategy prescribes how to play in the upcoming round—either attend, not-attend, or randomly select. Now, list strategies for all players as in Table 7.1, indicating what the "next play" is in light of pre-play and attendance data found in a row. Keeping things simple, we restrict the number of prior rounds considered to one. Relaxing this condition is one way to enhance strategic considerations. To illustrate, having attended in the previous round, S_9 forgoes swimming on the next turn, no matter the attendance and otherwise makes a random choice. If attendance was favorable in the previous round, S_8 picks at random for the next and does not attend the upcoming session if an unfavorable number turned up at the last.

Table 7.1 Strategies for the Swimming Pool Problem.

$0 \to$ not attend	$- \to\, > 5$
$1 \to$ attend	$+ \to\, \leq 5$
$R \to$ pick 0 or 1 at random	

Strategy	Pre-play	Attendance	Next play
1	0	+	R
	0	-	R
	1	+	R
	1	-	R
2	0	+	1
	0	-	0
	1	+	1
	1	-	0
3	0	+	0
	0	-	1
	1	+	0
	1	-	1
4	0	+	1
	0	-	0
	1	+	0
	1	-	1
5	0	+	1
	0	-	1
	1	+	0
	1	-	0
6	0	+	0
	0	-	0
	1	+	1
	1	-	0
7	0	+	1
	0	-	1
	1	+	0
	1	-	1
8	0	+	R
	0	-	0
	1	+	R
	1	-	0
9	0	+	R
	0	-	R
	1	+	0
	1	-	0
10	0	+	1
	0	-	0
	1	+	R
	1	-	R

With these strategies dictating players' choices, we can compute what happens over eleven rounds of play. Table 7.2 displays the results. In round zero—the initial condition, plays are randomly generated. Some of the outcomes arising from this experiment in *algorithmic choice* are clear. For the most part, attendance is close to the threshold value of five and is favorable more often than not—seven times. Some strategies produce attendance at a small number of agreeable sessions. In the extreme, S_6 never attends.

One response to these data involves modifying strategies so that more desirable circumstances occur. But, what modifications will ameliorate the results and how do you find them? You could tinker with the strategies, making small alterations and looking for improvement over several series of rounds. Testing a strategy-set by altering the values in round zero is a way of checking for stability. If the series produces significant change when the initial condition undergoes a small deviation, it's a good guess that the set is not close to optimal. In the context of dynamics, such behavior recalls the chaotic property known as sensitivity.

Speaking of dynamical systems, repeated application of the strategies amounts to iteration of a map that turns one round's values into those of the next. This comparison suggests that something interesting might take place if the iteration makes a long run. Since only a finite number of different rows can appear, the initial round's dynamical orbit must eventually fall into a cycle of some length—a fact that draws a stark contrast to dynamics on a space with infinitely many points. Is that cycle significant in some way? What if it turns out to be a fixed point? Using dynamics allows a choose-first approach to what is fundamentally a data-mining problem.

What's important to grasp is that the "object" being targeted for enhancement is the *collection* of strategies, not any individual strategy. How challenging is it to achieve this goal? Knowing the total number of strategies would help, as that would give us a sense of what the players have to work with. Any particular strategy is defined by selecting one of three plays for each of four states, meaning that there's a "deck" of $81 = 3^4$ individual rules S_1 to S_{81}. We want to know how many *distinct clusters* of ten strategies there are. Assigning one of 81 rules to each of ten players, we get $81^{10} \approx 10^{19}$ different *strategy-sets* that are "points" in a *strategy-set space*

$$X = \{SS_1, \ldots, SS_{81^{10}}\}.$$

Were there 100 participants, a *combinatorial explosion* produces $81^{100} \approx 10^{191}$ distinct collections containing 100 rules. Working in spaces this huge precludes conducting exhaustive searches for an optimal strategy-set.

Looking more deeply into the problem, how does symmetry get into the game? We have a space X consisting of 81^{10} points each of which is a set of ten strategies. When the deck of 81 rules is shuffled, certain strategy-sets can emerge unchanged. Symmetry! For example, take the set that contains strategies S_1 to S_{10} and let T shuffle the points so that

$$S_1 \to S_4 \to S_9 \to S_3 \to S_1 \quad S_2 \to S_6 \to S_7 \to S_2 \quad S_8 \leftrightarrow S_{10} \quad S_5 \leftrightarrow S_5.$$

Table 7.2 Strategy-driven Plays for the Swimming Problem. Values in the bottom row indicate the number of times that a swimmer found favorable conditions.

Strategy → ↓ Round ↓	1	2	3	4	5	6	7	8	9	10	att.
0	1	1	0	1	0	0	0	1	1	1	6
1	1	0	1	1	1	0	1	0	0	0	5
2	0	1	0	0	0	0	0	0	0	1	2
3	1	1	0	1	1	0	1	0	1	0	6
4	1	0	1	1	0	0	1	0	0	0	4
5	1	1	0	0	1	0	0	0	0	1	4
6	0	1	0	1	0	0	1	1	1	1	6
7	0	0	1	1	1	0	1	0	0	0	4
8	0	1	0	0	0	0	0	0	1	1	3
9	1	1	0	1	1	0	1	1	0	1	7
10	1	0	1	1	0	0	1	0	1	1	6
11	1	0	1	1	1	0	1	0	0	0	5
# favorable	4	3	4	4	4	0	4	0	1	3	

Whatever happens to the 71 complementary strategies

$$S_{11}, \ldots, S_{81}$$

is irrelevant. An apt term for this type of transformation is *combinatorial symmetry*.[1] Given the vast size of strategy-set space, the combinatorial symmetry groups can likewise be enormous, a fact that alerts us to the level of difficulty involved when attempting to optimize strategy-sets. Being precise, there are 10! ways of shuffling ten elements in a strategy-set among themselves and, for each of these, there are 71! ways of rearranging the strategies outside of these ten. Thus, the total number of shuffles that preserve a typical strategy-set is

$$10! \; 71! \approx 10^{108}.$$

Our discussion of the swimming pool problem now finishes by examining a couple ways in which stochastic processes can influence the interactions. These cases, at the least, make for worthwhile comparison to others. In the first instance, start with a collection of strategies whose plays are chosen at random. That is, every player determines a play for each of the four states by selecting 0, 1, or R according to some randomization scheme—such as a 25% chance for 0 and for 1, while R turns up with a likelihood of 50%. Then run the resulting strategy-set and look for beneficial tweaks. Instead of searching for improvement by inspection and intuition, the problem is amenable to a machine-learning algorithm the motivation for which stems from biological evolution. A later section will discuss in some detail how to implement an evolutionary process when faced with a tough choice of a certain kind.

---

[1] Combinatorics is the study of properties and functions associated with arrangements of objects.

As for the second technique, every strategy makes a random play for all states so that every member of a strategy-set plays R regardless of what state it encounters (see S_1 in Table 7.1). Although it appears that these conditions would make for the wildest sort of behavior, the overall—rather than individual—outcomes with respect to attendance are predictable with high accuracy, provided that a large number of rounds are played. Say that each strategy plays 0 and 1 with a 50% chance. On average, you expect five zeros and five ones to be played in a round. Of course, the exact numbers will vary around this mean value. Elementary probability tells us that swimming conditions will, over many rounds, be favorable (attendance less than six) about 60% of the time. Since each strategy is equivalent to any one of the others, participants will get to swim in favorable conditions about the same number of times. This seems to be a practice that's equitable and easy to implement. An interesting query is whether a better strategy-set exists, something that requires clarification as to what 'better' means.

On a historical note, the swimming pool problem is a restatement of a conundrum known as the El Farol problem that's associated with the Sante Fe Institute. Patrons of the El Farol bar try to adopt certain strategies in order to avoid going there when it's uncomfortably crowded.

7.2   THE HEART HAS ITS REASONS

Consider a couple in the market to purchase a house. One partner tells the other about two new prospects. Using a bit of moral algebra to examine all of the properties' important attributes, they aren't left with a preference for one house over the other. However, after visiting both places, one party clearly prefers a particular house. What triggered the change of opinion were the house numbers: 846 and 1313. It turns out that the partner in question is triskaidekaphobic, a condition that settles the matter in favor of the house at 846.

This vignette points to a common reason for choosing among options that otherwise are assessed equally, namely, some form of bias. Without being aware of it, we're capable of carrying around predilections and suppositions that influence our beliefs and thereby the processes by which we make choices. While we often think of action due to bias as harmful, assumptions are essential to our experience of and interactions with the world. Who behaves according to a blank slate of ideas and values? What would it even mean to do so? In a way, biased thinking can act like a computational tool that nudges or pushes us in a certain direction.

Next, without passing judgment regarding harm versus benefit, let's survey a few instances of "bias algorithms" that guide deliberative processes—not the same thing as biased algorithms. While we might not be able to do without biases altogether, understanding something about their provenance and how they operate can help guard against their baleful influence.

Call the first example *chirality bias*, which amounts to favoring right over left or vice versa. Note that this partiality can be taken literally—referring to right and left hands—or figuratively as in political persuasion. If Buridan's mule is left-handed (left-hoofed), it bypasses the balance between hay bales accordingly. Faced with two equally qualified electoral candidates, a voter who identifies as right wing selects the one whose party affiliation aligns more closely with that preference. An ambidextrous mule has a more difficult choice as does a centrist voter. A price paid.

While the extent to which we choose our beliefs is debatable, what we believe clearly has an influence on our choices. To the algorithmic mindset that we've adopted, a system of beliefs serves a computational purpose when it comes to making decisions. A particularly sharp instance of this dynamic occurs when someone embraces a conspiratorial view regarding some aspect of the world. With respect to their effects, some conspiracy theories are insignificant whereas others are consequential.

Taking a current affair, the latter category surely includes the constellation of ideas, opinions, and values that nourish rejection of the position that human industry is altering climate in ways that call for aggressive measures. What kind of bias might promote a point of view so deeply at odds with a broad and multidisciplinary scientific consensus? Perhaps there are multiple possibilities. For one, take an extreme libertarian line of reasoning that opposes governmental regulation as a matter of course. The thought process might go like this. Actual large-scale climate disruption would justify robust intervention by states across the earth. Since action at such a level is at odds with basic values, a credible alternative narrative must be found. A story that answers this need is lifted from an account in the annals of the "dispute" about smoking's ill effects. Namely, there are grounds to doubt the claims made by scientists versed in various aspects of the subject [6]. Employing a choose-first practice can provide an antidote to this and many other *conspiracy biases*. An appropriate later section will address this point.

Succession bias occurs when someone is disposed to conclude that an earlier event is the cause of one that succeeds it. As a type of correlation-implies-causation mistake, the sentiment that rises when events are conjoined can be quite compelling. After dining at a new restaurant, you experience insomnia that night. You then claim that the meal is responsible for the physical distress and decide to avoid the establishment. A headache disappears after taking medication, convincing you that the pain ended because of the medicine. It could be, but also might not be. Emotion can run high in such circumstances. When a child starts to show signs of autism shortly after taking a vaccine, how great a kneejerk temptation it must be for a parent to blame the pharmaceutical.

For each of these effects, there are other plausible causes that are roughly indistinguishable according to a moral algebra calculation, say. In the autism example, there's genetic propensity, an illness experienced prior to the onset of symptoms, or the family's move to a new house before autistic behavior began. Succession bias alone does not account for which preceding condition achieves favored status as the actual cause, and so, fails to break the (shuffle) symmetry among the possible causal events. Looking to refereed studies that address the viability of a plausible cause is a rational way to decide. What could also do the trick is intervention by an additional bias, which we take up next.

Seizing on a conspiracy story can incline someone toward a particular causal explanation. Accepting that public health agencies alongside pharmaceutical enterprises are deceiving a society about vaccine safety and efficacy gives credence to the claim that vaccination and autism can be causally connected. Once a conspiratorial picture comes into view, a follower is prone to see verification for it regardless of what circumstances obtain. On one hand, a school's vaccine requirement is a clear sign of the state's complicity in boosting profits for the manufacturer. On the other, if no mandate is enacted, there must be a cover-up underway.

A pair of biases that can affect our beliefs and consequently our choices are related by virtue of being opposites of a sort. Person P falls under the sway of *conformity bias* when P tends to believe (reject) an assertion when many of those in P's circle of acquaintances believe (reject) the claim in question. How many actors it takes to attain a quorum that moves P to follow the crowd depends on context. Important parameters that govern P's response are the frequency with which and order in which P encounters crowd members. Of course, large-scale social networks can provide a potent force on thought. Conformism of this sort might well be a means by which conspiracy thinking takes hold in an individual.

The flip side of conformity bias is that of *rebellion* according to which someone rejects (adopts) an opinion held (not held) by a substantial proportion of folks in contact with that person. A child who adopts certain beliefs simply because those positions are contrary to parental views provides a classic example. In spite of being the polar opposite of conformity, rebellion bias can also push someone towards acceptance of the tenets offered by a conspiracy theory. This polarity speaks to the Frost poem quoted at the beginning. Conformity selects the road *more* traveled whereas rebellion chooses the one less traveled.

A kind of propaganda that can have pernicious effects relies on repeated utterance of a message. Someone whose belief system is susceptible to this kind of signal suffers from *repetition bias*. Manipulation of this sort can be especially harmful when the message is false or lacks evidential support. Political campaigns are common sources of such activity.

Another tactic sometimes employed by candidates for office and by commercial advertisements involves telling a story that's too good to be true and yet proves to be convincing for many in a target audience. Being taken in by such a fantasy signifies a case of *fairy tale bias*—a kind of motivated reasoning whereby someone holds onto a belief due to a strong desire for it to be true. Think of someone who, in the absence of evidence, takes an herbal supplement or off-label medication primarily because they want the treatment to work.

Or of a voter whose support for a candidate making unrealistic promises is largely due to a wish that the promised results will materialize somehow.

In our limited survey, we can recognize that overlap and interplay can occur among various biases. Someone might see successive events as causally related out a desire for them to be so connected. Moreover, that fairy tale belief could stem from a propensity to attribute conspiratorial behavior to some entity. For instance, someone wants to see a causal connection between vaccination and autism due to their belief in conspiracy within public institutions.

At a general level, a sense of infallibility can develop in and overwhelm the thought processes of a biased mind. Such an outlook is something of a meta-bias whereby someone holds to an opinion despite lack of evidence for it or worse, evidence to the contrary. What might happen here is that one type of bias supports another. For an economic example, some policy makers and even some economists take the view that lowering taxes on wealthy people is an action that pays for itself by stimulating growth. There's abundant research that gainsays the claim and little in its favor. Fairy tale bias could account for why these folks hold this position while infallibility bias—the tax-cut claim just can't be wrong—justifies belief in the fairy tale.

We often view states of mind under the influence of biases as harmful—when leading to types of chauvinism, say. They can, however, be natural, albeit hidden, aspects of our experience. Indeed, the only means to avoid making suppositions is to choose randomly in all circumstances, hardly a tenable form of life. Let's close the discussion with some examples of biased behavior that seems benign.

In some influential psychology experiments, Kahneman and Tversky revealed a common human bias—one that you might find surprising [26, 35]. Call it *loss-over-gain* since it has to do with subjects' having a preference for avoiding loss over achieving gain, even when the two outcomes are equivalent in some sense. To make the discussion concrete, consider two situations in each of which someone is offered a choice:

1	A:	receive $100	2	C:	lose $100
	B:	50% chance to win $200		D:	50% chance to lose $200

For a given situation, each option has the same expected value—a probabilistic measure where an outcome is multiplied by its likelihood. For scenario 1, the expected value of A is $100 since its probability is 1, while B's is

$$\$100 = \$200 \cdot (.5).$$

Similarly, expected values for the second choice are $-\$100$ and

$$-\$100 = -\$200 \cdot (.5).$$

Relative to this property, A and B are equivalent as are C and D. In other words, there's an exchange symmetry between the options in each case. However, most subjects select A—the sure thing—when presented with the chance to gain and *D*—the gamble—when loss is on the table. These empirical results indicate that human psychology values loss-avoidance over gain-acquisition, rejecting the notion that the two states are related symmetrically.

A composer begins writing a piece of music with a single note—perhaps one that appears first in a number of scores. Next comes a decision regarding which note follows the first and then what the third note is to be, and so on. Putting the first note in place restricts what the second can be and having those two in place constrains the artist to a greater extent when choosing a third note. These constraints emerge from a complex web of aesthetic valuation and proclivities that serve as a guide to the creative process. That is to say, a kind of artistic bias guides the composer's craft. Stringing together "elementary musical particles" is not a deterministic activity. Otherwise, the composer could produce only a small number of scores. For example, the first few notes of two pieces might be identical until a "mutation" causes the second to diverge from the first. Our upcoming examination of evolutionary systems will explain how random mutations are a kind of symmetry-breaking feature. Whether or not our composer makes random choices could be a hard question, but such a procedure can be incorporated into a music-composing algorithm that selects some notes according to a stochastic scheme. Say, pick C with 50% probability, D with 30%, and E with 20%.

Music composition bears comparison to playing a game of chess. Suppose the moves made in every game up to now have been recorded in a book of games. For some relatively small number of moves, any particular game is identical to at least one in the book. If a game goes long enough, there will appear a crucial move that first differentiates the current sequence from *all* that have gone before. This assertion follows from the combinatorial explosion in the number of games that can be played using a few hundred moves. The threshold beyond which the game is no longer repetitious has a mutative quality. Among skilled chess practitioners, moving pieces at random seems an unlikely tactic whereas musical invention might allow a bit of randomness into the creative act.

7.3 GIVE CHANCE A CHOICE

An important theme that runs through much of this work concerns the role that random action can play in decision-making, due to its propensity for breaking whatever symmetry makes a choice onerous. The current section touches upon several instances in which chance is built into the computational means by which some choices are made. In doing so, we extend the notion of choosing beyond the standard account of an action performed by a conscious agent. In effect, our concept of an agent will expand.

Taking a risk

Mathematical thought is a suitable setting for the notion of a *pure choice*, one that presents truly indistinguishable options. However, conditions aren't quite as pristine in physical and social worlds, where imagination is not the governing precept. Consider the matter of assessing risk—a task with which many humans struggle. And yet reasonable estimates of the risk involved in making choices are clearly valuable. The first issue is to clarify just what we mean by risk. One interpretation commonly adopted construes the risk associated with an event as tantamount to that event's likelihood. A more perceptive account ties the probability of an outcome to its value or utility according to some measure.

Gambling often provides seminal examples of topics in probability. To play the carnival game *Over-Under* you make a bet that a pair of thrown dice will turn up over or under seven. Guess correctly and you win $3. Otherwise, you lose $2. One kind of question that probability theory asks is how much you can expect to win or lose, if you play many times. The analysis that follows does not apply to just one or a few bets. There are 36 possible outcomes for each throw. Of those, 15 are over seven, 15 are under, and six are equal to seven. Say you bet on over each time. In the long run, you win $3 on 15 out of every 36 throws and lose $2 on 21 out every 36 throws. So, for each roll of the dice, you can expect to gain

$$\$3\frac{15}{36} = \$\frac{15}{12}$$

and lose

$$\$2\frac{21}{36} = \$\frac{14}{12}.$$

Rather than taking into account only the chances of gain versus loss, a better test compares expected gain against expected loss. The difference between these two quantities is a measure of the risk you're taking. For Over-Under, this quantity amounts to $1/12 which is a favorable risk despite there being a greater chance of losing on any particular roll. Of course, no self-respecting "house" would host such a game. This example serves merely an expository purpose.

When making a binary choice between A and B, a judgment based on a calculation of risk weighs two quantities against each other:

$$\text{value}(A) \cdot \text{prob}(A) \qquad \text{and} \qquad \text{value}(B) \cdot \text{prob}(B).$$

In many situations, assigning a probability to an event lacks precision and might be handled better with a Bayesian treatment. As we discuss in another context, doing so involves adjusting probabilities that are conditioned on updated circumstances.

For another illustration of how risk calculation can influence a decision process, suppose you're driving on a winding mountain road and come across a slow-moving vehicle. At no point can you see far enough ahead to be assured that you can pass safely. But, there has been very little traffic on the road. Do you attempt to pass or not? First compute the expected gain and loss if you decide to pass:

$$\text{gain: value(no-oncoming-car)} \cdot \text{prob(no-oncoming-car)}$$

and

$$\text{loss: value(oncoming-car)} \cdot \text{prob(oncoming-car)}.$$

The chance that a vehicle occupies the other lane is low so that prob(no-oncoming-car) is close to one. (A probability is a number between zero and one.) Passing successfully saves a time delay whose measure is captured in value(no-oncoming-car) as a finite number. In the unlikely event that a car is oncoming, assigning an infinite value to the outcome seems reasonable. The result is an expected loss of $-\infty \cdot$ prob(oncoming-car), a quantity equal to negative infinity no matter how small the chance of an oncoming car happens to be. So long as prob(oncoming-car) is not zero—which it isn't, the risk of passing is always unfavorable.

Selective evolution

There are many ways for a system of some kind to evolve over time. Random processes can be a key mechanism that drives such change. Here, we treat stochastically-driven evolutionary systems as a compute-first means of choosing outcomes. In an ecological setting, the object that experiences evolutionary change is a collection of organisms—a species. Driving the development are processes that introduce variation into a population, such as gene combination and mutation during genetic replication. Each mechanism functions with a degree of randomness. The way that genes combine after mating is sensitive to environmental conditions—recall the properties of dynamical chaos. Where and when mutations occur as genes get copied are unpredictable events, perhaps inherently so.

These methods by which a population transforms are analogous to computational operations. Pushing the analogy further, we can regard the transformational genetic activity as a kind of symmetry in which the difference between original and resulting phenotypes appears to be undetectable. Our analytic framework calls for a choice that evades the symmetry. Natural selection provides a mechanism through which an ecosystem "chooses" a phenotype with greater fitness, provided that suitable traits arise. Ultimately, a novel species can emerge after a sequence of choices due to ecological pressure. Of course, indiscernibility among competing phenotypes is not necessary in order for natural selection to operate.

In the chapter that follows this one, we examine a computational system that stochastically evolves through selection. The algorithm there admits a choose-first interpretation.

Routine teaching

Much of human knowledge and behavior—perhaps non-human as well—depends on classifying things. Working out a classification scheme amounts to choosing proper categories into which objects are placed. For example, in a given situation what actions are safe, dangerous, or neutral? To what degree? Categorizing members of a set is a task for which machine algorithms are sought.

Say that you want an application that visually recognizes flowers. To keep the discussion simple, the device should be able to distinguish flower from non-flower. A standard artificial intelligence approach to a problem like this is to develop a procedure through training. Beginning from a null state, the program "sees" a long succession of images—the *learning data*—each labeled flower or non-flower. Over time, the goal is for the device to pick up on properties shared by flowers but not possessed by non-flowers. Arriving at an ability to perfectly discriminate between the two categories might not be easily accomplished. For instance, discerning characteristics seen in flowers but not in fan-blades could be somewhat tricky. The difficulty amplifies with the task's subtlety. Facial and voice recognition systems must acquire the ability to detect fine differences. There's also a danger that the training data allows preferential bias to make its way into an algorithm's output [28]. Even the sequence in which a machine encounters the material on which it learns can affect its performance due to an order-effect bias. A chess-playing program classifies a vast number of possible moves for a given board configuration as a route to learning which maneuver is optimal or close to it.

Empirically trained algorithms provide a computational means to an end whereby classificatory judgments can take place. A role for randomization can be found in the training regimen, inasmuch as arranging elements of the learning data in a haphazard way can mitigate the prospect of order effects as well as other forms of bias.

CHAPTER 8

Choosing to Learn

For solutions to polynomial equations based on iteration, there's a theoretical key that unlocks the barrier erected by symmetry: the connection between a dynamical system that respects a realization of a polynomial's symmetries and an act of choosing that does not. In every instance, you locate a special set of points using an iterative computation *after* making a symmetry-breaking choice (of initial condition). The current chapter's itinerary visits an assortment of settings in which you can arrive at a decision by first choosing and then computing. Such a *choose-by-choosing* method might sound paradoxical or vacuous, but not when viewed in light of iterative solutions to equations.

8.1 A CROWD DECIDES

In a population, choices can take place at both individual and aggregate levels. Here we give various illustrations of the interplay between these two types. Sometimes a collective decision can possess properties not shared by any of the choices made by individuals.

Averaging choice

A popular elementary school activity presents a class with a jar full of beads. The container is transparent and the children are asked to guess how many beads it contains. They can inspect the closed jar only by sight and there are far too many beads to be counted by simple enumeration. One-by-one the kids report their estimates after which the teacher reveals the actual number. In this case, it happens that the student who goes last gives a value quite close to the true one, besting the other guesses. Some questioning reveals that the best guesser computed the average of all other guesses. Noting that the class contains 27 members and calling their conjectures G_1 to G_{27},

$$\text{revised } G_{27} = \frac{G_1 + G_2 + \cdots + G_{27}}{27}.$$

To clarify, the 27th student makes an initial guess and then revises it by calculating the class average.

The clever student shows how a *collective choice*—in this case, estimating—can be superior to those made by individuals. More generally, this phenomenon illustrates a way in which choosing emerges at the population level. Accounting for this crowd effect is not difficult. Each student looks at the jar and comes up with some way of estimating the number of beads. For any guess that overestimates the exact quantity by some amount, it's likely that another guess will underestimate the number by roughly the same amount. Averaging the two guesses gets closer to the correct value than either individual estimate. Increasing the population typically yields more accuracy in the average as values spread out in a more balanced arrangement.

Swimming with Bayes

Revisiting the swimming pool problem introduced in Section 7.1, an alternative tool that's applicable to a swimmer's decision to attend a session or not involves conditional probability. Specifically, swimmers want to know how likely it is for attendance to be favorable given that they go to the pool. Bayes's rule states that this conditional chance is equal to the product of 1) the probability that the swimmer attends given that attendance is favorable and 2) the quotient of the chance that attendance is favorable and the probability that the swimmer attends. Expressed verbally, the rule is a mouthful. A formal equation captures the relationship between the various probabilities more succinctly. Designate by F and A the states that attendance is favorable and the swimmer attends:

$$\text{prob}(F|A) = \text{prob}(A|F)\frac{\text{prob}(F)}{\text{prob}(A)}.$$

Read the vertical slash as "given that." It makes sense that there's a strong chance that a swimmer attends if attendance is favorable; that is, $\text{prob}(A|F)$ is close to one. If we set it equal to one, the formula becomes

$$\text{prob}(F|A) = \frac{\text{prob}(F)}{\text{prob}(A)}.$$

On the right side, $\text{prob}(F)$ can be estimated by observing attendance numbers while $\text{prob}(A)$ results from a swimmer's choices. According to Bayes a swimmer should moderate their attendance, since a high value for $\text{prob}(A)$ tends to decrease $\text{prob}(F|A)$. So, a Bayesian swimmer decides how frequently to go to the pool and then adjusts the frequency in light of updated information regarding attendance history. With that, using conditional probability to guide the choice of when to swim incorporates a choose-first procedure. Letting the "population of choices" play out leads to a refinement in decision-making, amounting to a computational component.

Transitivity: A test of choice

A prominent type of collective decision-making occurs in an election where individual choices produce an outcome for a community. When a contest is between two candidates, things are simple: the candidate who receives the majority of votes is preferred by a majority of voters. Including a third candidate can create perplexing scenarios. A three-candidate voting phenomenon exemplifies a principle found elsewhere. For instance, when passing from a gravitational system consisting of two bodies to one that contains three — such as from a star with one orbiting planet to a star with two planets, the dynamical characteristics of the respective systems can undergo qualitative change. A two-body interaction tends to a stable configuration regardless of initial state. When a third mass is involved, the system typically develops an unstable trajectory. Let's illustrate such an alteration in a three-candidate election.

In our simplified election, there are seven voters and candidates A, B, and C. The election is conducted "head-to-head," meaning that the overall winner is the candidate who wins the most contests between pairs. First notice that in two-candidate head-to-head voting, whoever receives a majority wins. Table 8.1 shows, along the rows, voter's preferences in each pairing with preference going to the candidate listed on top. For example, voter 1 prefers A to B, A to C, and B to C. Results in the columns determine total preferences and thus, the overall winner. In the bottom row, pairwise preferences show A favored over B and B over C. Invoking the property known as *transitivity*, it seems reasonable to expect that A would be preferred to C. For instance, "less than" is a transitive operation on numbers:

$$\text{If } x < y \text{ and } y < z, \text{ then } x < z.$$

However, such is not the case here since C tops A. What's more, votes cast by individuals adhere to the principle of transitivity, an outcome consistent with rational behavior. So it seems that rationality at the micro-level can produce macro-level irrationality. Perhaps we should take this circumstance as a sign that head-to-head voting is a faulty system.

Analysis of voting systems is a rich and subtle field of mathematical study [3]. Our toy example is intended to showcase how a collective decision can exhibit a property shared with none of the individual choices that contributed to it.

Consider another situation related to deciding an issue where a failure of transitivity occurs. A family in the market for a house has found three attractive prospects. In the spirit of prudential algebra, they place weights on three categories: price (P), school quality (S), and work proximity (W). Labeling the houses 1, 2, and 3, it happens that

$$P_3 > P_2 > P_1 \qquad S_2 > S_3 > S_1 \qquad W_3 > W_1 > W_2.$$

To avoid confusion, note that these weights indicate how much value is attached to a category for a given house. For instance, $P_3 > P_2$ means that

Table 8.1 Three-candidate Election.

Voter	A vs. B	A vs. C	B vs. C
1	A B	A C	B C
2	B A	A C	B C
3	A B	C A	C B
4	B A	C A	B C
5	A B	C A	C B
6	B A	C A	B C
7	A B	A C	B C
(total)	$\frac{A}{B}$ (4)	$\frac{C}{A}$ (4)	$\frac{B}{C}$ (5)
(total)	$\frac{B}{A}$ (3)	$\frac{A}{C}$ (3)	$\frac{C}{B}$ (2)

house number 3 is more favorable on price than is number 2; that is, the price of 3 is less than the price of 2. When they consider price, the valuation shows 3 besting 1. For distance to work, 1 beats 2 and 2 tops 3 on school quality. Thus it happens that the overall house ranking lacks the transitive property that would make the family's decision more clear cut. The favored house would have been 3 if its schools had a rating higher than those associated with 2. When choosing among ranked options, the absence of transitivity can hinder the deliberative process, calling to mind the way in which symmetry throws up an obstruction to solving equations. We can portray this sort of symmetry by arranging the preferences in columns and noting that the bottom row is a cycled version of the top.

Table 8.2 Symmetry Associated with a Failure of Transitivity.

	P	W	S
	3	1	2
preferred to	↓	↓	↓
	1	2	3

Choosing can be contagious

An economy is many things, a basic component of which involves aggregating individual choices into a collective one. This perspective is central to behavioral economics, a field that studies, among other things, how large-scale impacts can arise from minor tweaks to behavior among economic agents, be they consumers, producers, traders, etc. A market is an arena where assorted agents interact. One of the main purposes realized by market activity is to set a price on goods and services. Associated with such costs is inflation—the rate at which prices change with respect to time. There are various ways of characterizing this important quantity. For our discussion here, we consider a somewhat nebulous "headline" inflation: the overall rate for an assortment of items familiar to the general public.

Numerous factors that are related in complex ways can drive inflation, making its behavior difficult to comprehend and forecast. An especially significant issue concerns the likelihood that inflation will change and how fast—a second order effect since the quantity is already a rate. Reference to the "inflation rate" can be misleading. Often what's meant is inflation itself and not the rate at which inflation is changing.

Of particular interest to us is a dynamical process that can cause both prices and inflation to increase. Suppose a number of consumers become concerned that prices for a "basket" of commodities will rise in the near future— that inflation is positive. In order to avoid the anticipated higher cost, they individually decide to purchase more of the goods than they would otherwise. If there are enough of them, their actions can push demand for these items substantially higher. Assuming that the supply of the items is unchanged, the increase in demand will boost prices. A dangerous feedback cycle looms: fear of inflation leads to actual inflation which further intensifies inflationary fear that leads to more inflation, and so on. Should demand accelerate—that is, increase at a rate which is itself increasing, the inflation rate will also grow.

Bearing a likeness to attraction in a map, the inflation-cycle can draw in more participants who worry that prices will soon rise. Enlarging the crowd can strengthen the effect, pushing the system toward a phase transition in which there's a critical population threshold below which the cycle fades and above which it takes off. In the latter context, we again see collective properties emerging from choices first made at the level of interacting agents. The dynamic here is not unlike what occurs in a forest fire where the density of trees—analogous to the ways in which inflation-phobes are distributed and receive information—influences how much of the forest burns. Here as well, a critical density marks the boundary between a fire either burning itself out or spreading across much of a landscape.

Another type of dispersive process along the lines of spreading inflation fear or tree combustion is an *information cascade*. Such a phenomenon occurs as a dynamism of choices passing from agent to agent. For a contrived example, say that a couple friends are out looking for a restaurant and happen upon two that are side-by-side. The types of cuisine are equally acceptable. Which do they choose? Their decision isn't quite Buridanic though; one place is empty and the other has one table occupied. That's enough to tip the balance in favor of the restaurant with customers. After the second group is seated, a third party arrives also with no partiality toward either style of food. But, they reason that the cooking must be better at the patronized location. As more parties appear, the one establishment fills while the other remains vacant. At each step in this process the arriving party extracts information from the circumstances they encounter. The crucial choice belongs to the first party, one which breaks the symmetry between two empty restaurants and sets off a cascade of subsequent choices. A heavily skewed collective decision ensues. Is it the case that seeing just one group dining somewhere justifies a belief that its dishes are superior? Maybe the first-arriving party prefers what's served at the restaurant they chose. Perhaps their position was akin to the mule's and they selected more-or-less at random.

The sort of information transfer that takes place in the cases touched on is also seen in colony behavior. Consider a cluster of foraging ants. As one member moves along a twig that splits into two, it "decides" to take the right branch. When a second ant approaches the fork, it checks both branches and then follows the chemical signal left by its predecessor along the rightward path. A third forager discovers the split and keeps to the pheromone trail made by the first and second ants. Just as consumers act collectively to increase their consumption and arriving diners make a group choice of one restaurant, the pheromonal information cascade leads to a steady stream of ants taking the route more traveled.

8.2 WHEN IN DOUBT, SIMULATE

Some of science's most effective tools are *models* that capture aspects of the way a system of interest behaves. Two types of model dominate the undertaking. First, going back to Newton at least, a model can be a set of equations that govern a system's variable elements. A more recent development, due to the advent of widespread high-performance computing technology, are *agent-based models* in which a collection of individuals interact according to rules that are sensitive to parameter values. Such variables describe certain features of a system such as behavioral, dynamical, and technological dependencies.

In our exploration of algebraic methods that uncover roots of polynomials, we worked with equation-based modeling. The alternative approach to solving equations—one that relies on discrete dynamics—exhibits a blending of the two kinds of model. In addition to formal expressions associated with maps, there is a dynamical push and pull among points, things we can regard as agents in a complex system.

This section and the one that follows look into a variety of models in which system features emerge due to agents interacting through their choices. A typical means of realizing such interplay is by choosing specific values for a model's parameters. Afterward, running the model gives rise to a simulation in which interactions among agents provide a computational service. In many situations, the parameters will be subject to dynamical change, meaning that the result of a simulation leads to an adjustment in the model itself. A good example of model evolution like this occurs in simulations of climate phenomena.

Turning a model into a simulation calls for assigning values to variable parameters that are incorporated into equations or interaction rules. When a dynamical system arises from a set of equations, initial conditions act as a kind of parameter. Once a user chooses these optional quantities, the model can run, generating either a continuous or discrete orbit. Although it's built on some amount of idealization—often by neglecting certain parameters, a model can still reveal significant features of the actual system that it's attempting to realize, at least partially. We saw how a dynamical orbit can visit a sequence of locations as it explores a space of choices, simulating a "machine" that yields an exceptional output—such as a periodic cycle of critical points. Next, we sketch several examples of simulations driven by agents interacting in a choose-first regime.

Strategic choice

Games provide natural media in which to run simulations. Our game theory treatment of the swimming pool problem views the adoption of a strategy as a computational tool. Alternatively, you can think of it as a choice whose virtue is disclosed by repetitive play against the strategies taken up by the other participants.

Many games admit this sort of recursive play. In some—such as Rock-Paper-Scissors, no single play is superior to any other. With repeated rounds of some games, certain strategies might work better than others. Two well-known examples that are a bit more subtle are Prisoner's Dilemma and Hawk versus Dove. To set up the former, a pair of colluding criminals, A and B, are arrested and each is offered the same deal as spelled out in the payoff structure given in Table 8.3. The "points" indicate how long of a prison sentence they can expect for each possible outcome. A prisoner betrays or cooperates by incriminating or not their accomplice. In Hawk versus Dove, players choose either to launch a military attack on an opponent (hawk) or to negotiate a peaceful resolution (dove). Numerical scores indicate how much damage to each side results from their choice of plays.

We can regard the playing of these games as either compute-first or choose-first operations. Before deciding which way to go, prisoner A can "calculate" that betrayal is the best play—simply by noting that whichever course B takes, A's best play is to sell out B. The trouble with this reasoning is that B sees things the same way. Rationality leads each prisoner to betray the other, giving both a sentence longer than what they would get were they both cooperative. That the payoffs lead each prisoner to the same play is a kind of symmetry in the game, crucial for the dilemma's appearance. In other words, the prisoners can trade places and the game is unchanged. Notice that Hawk-Dove does not suffer from this sort of difficulty. If A is hawkish, B's best play is also as a hawk. But, if A is a dove, B is best served by playing dove as well. From B's point of view, the same implications hold since the Hawk-Dove payoffs are unchanged if A and B trade places. Perhaps it's not difficult to believe that Hawk-Dove can be enriched and adapted to war games—simulations made all the more important in an age of nuclear weapons.

Table 8.3 Payoffs for Prisoner's Dilemma and Hawk vs. Dove.

	$A\downarrow$ $B\rightarrow$	Cooperate	Betray
Prisoner's dilemma	**Cooperate**	$A:-2$ $B:-2$	$A:-5$ $B:0$
	Betray	$A:0$ $B:-5$	$A:-4$ $B:-4$

	$A\downarrow$ $B\rightarrow$	**Hawk**	**Dove**
Hawk vs. Dove	**Hawk**	$A:-10$ $B:-10$	$A:-5$ $B:-15$
	Dove	$A:-15$ $B:-5$	$A:0$ $B:0$

Either of these games, and many others, can run as choose-first systems by beginning with a choice of strategy for each player in a population. Members in the collection of strategies count as agents that interact by matching up according to some multi-layered scheme—being neighbors in a "play-space," say. Most simply, players engage one-on-one with others nearby in a temporal sequence of multiple rounds, using the same strategy in each round and against every opponent.

A richer mode of play further accommodates a variety of spatial settings covered by a grid of cells. For instance, basic configurations use rectangular cells on a plane or in space of any dimension. We can also imagine carving a cellular layout into other shapes—such as a sphere, donut, pretzel, and so on. Activity starts with each cell's selection of a strategy that dictates how it will play against its neighbors—characterized in some way. On a rectangular grid in the plane, a cell's neighborhood might be the four with which it shares an edge or the eight with which it shares a vertex. As an example, a cell's initial move could be to make a random play against each neighbor. Once a cell's strategic history exists, its interactions with neighbors can prescribe the strategy that it adopts in the next round. Out of the many algorithms to which a cell can be subject, here's one. Consider a cell C and one neighbor N. To decide what C's next play against N will be, select the highest scoring play between N and its neighbors, of which C is one. By iterating the procedure, we compute dynamical orbits in the space where all of the cells' strategies reside.

Playing out a conspiracy

In an earlier examination of compute-first phenomena, we construed conspiracy bias as a kind of calculation that influences decision-making. When concluding that discussion, we took note that a choose-first technique can also serve as a countermeasure to buying into such a narrative. The idea is to suppose that you accept the core statements of a conspiracy story—in our example, that climate-change science is a sham. Then run a conceptual simulation by asking what further opinions—likely hidden from scrutiny—does the conspiracy account commit a believer to holding. A key claim of this sort is that there must exist an arrangement among a vast international network of research scientists who work in disparate fields such as atmospheric physics and chemistry, oceanography, ecology, meteorology, and so on. They must have agreed to slant or fake their peer-reviewed papers in such a way that what they collectively report hangs together as a coherent body of knowledge while being subject to an airtight deception and secrecy. Of course, accepting that such collusion takes place is what turns a constellation of beliefs into a conspiracy theory. That such mendacity could be so thoroughgoing as to be executed without the appearance of a single whistle-blower or intelligence leak is preposterous.

A little bit of bias can go a long way

We can now add to our survey a form of bias that can display subtle features. *Homophily* occurs when each agent in a population has a degree of preference for other agents with whom it shares some property—such as species, ethnicity, race, education, income. A model of homophilic behavior grew from ideas of the economist Thomas Schelling [41]. It proposes that each agent chooses a threshold for the proportion or configuration of dissimilar agents that it tolerates as neighbors—however the notion of neighboring is characterized. In the simplest case, a simulation that implements a homophilic condition takes place on a cellular grid and every agent adopts the same threshold value. If the triggering measure is exceeded, the affected agent moves to another location according to some principle—for instance, nearest vacant cell or a randomly selected vacancy. Otherwise, the agent makes no move. A diagram in Figure 8.1 illustrates how the system acts. When the positions of all agents have been evaluated and the corresponding actions taken, a second round is carried out, then a third, and so on.

This process defines a dynamical system that's sensitive to random influence and capable of sudden jumps. Naturally, we want to know how a population of agents behaves after many iterations. One surprising finding is that it does not take a low threshold in order for agents to end up in a stable arrangement in which they segregate on the basis of their distinguishing property. That is, agents can exercise a relatively high tolerance for dissimilarity, but still tend toward segregating themselves. Arriving at such global stability seems to be helped by having a supply of empty locations adequate for "buffering" around clusters of similar agents. In our prior treatment, biases played a computational role. Here, in a case of role-reversal, bias is cast as a mechanism of choice.

X	O	O
X	O	O
O	O	X

Figure 8.1 Simulating homophily. In the simplest situation, there are two classes of agent, denoted X and O. The "world" in which the agents live is represented by cells in a grid, each agent with eight neighbors. Applied to the agent O in the center cell, homophilic bias causes it to move to a vacant cell if, say, three or more of its neighbors are Xs—as there are in the sample. With fewer than three agents of the other category, the central agent remains in place.

A choice problem

At a place where politics and mathematics meet, the design of U. S. congressional districts is a formidable undertaking [19]. Without diving into subtle details, we call attention to a choose-first method of sculpting the boundaries of a district map (or "d-map" so as not to confuse this usage with the one employed in our treatment of dynamics). The "districting problem" concerns optimization, at least in the abstract. Once you decide what makes one d-map *better* than another, determining that some d-map is optimal—better than every other configuration—is likely impossible to achieve. There's a vast infinity of possible d-maps for a state, many of which have a very high degree of complexity. Certain basic principles can be imposed on district maps—for example, individual districts should come in one piece. The most that can be hoped for is finding an arrangement that attains a certain level of fairness. So, we need to characterize what it takes for a d-map to be *fair enough*. Doing so is a significant hurdle that the districting problem presents. We now sketch a means to that end.

Assuming that there are two political parties A and B, one such standard takes the representation by each party arising from a given d-map and compares it to partisan representation across the state. For a particular configuration of districts and distribution of votes, say that party A wins 70% of the seats, but receives 50% of the statewide vote while B takes 30% of the seats and 50% of the total vote. Is this gerrymandering? What about a 60%/40% breakdown? Suppose we agree that a d-map M is not gerrymandered if the absolute value of the difference between seats won by a party and statewide votes received by that party is less than 5%. Call this quantity a gerrymander detection function (note the absolute value):

$$G(M) = |(\% \text{ seats won using } M) - (\% \text{ seats won statewide})|.$$

We can attack the problem dynamically. For an initial condition, *choose* an arbitrary d-map M_0. Making such a selection is not trivial given how enormous the space of d-maps is. Next, simulate the statewide vote and check the difference between seats and votes. If it's absolute value is greater than 5%, make a small change in the M_0 to obtain a new d-map M_1. Run the simulation on M_1 and test for gerrymandering. Iterating the procedure *computes* a dynamical orbit of d-maps. Stop if some d-map M_k passes the gerrymander test. To check for robustness in M_k, simulate the distribution of votes in multiple ways and see how the candidate d-map holds up to the gerrymander test. Without question, the most delicate bit in this algorithm concerns the alteration process from one d-map M_k to the next M_{k+1}. Ideally, we want a technique that will decrease G under a variety of vote distributions; that is,

$$G(M_{k+1}) < G(M_k).$$

This is a hard problem indeed, but one that fits our theoretical framework.

Choose the right thing

One type of choice that's basic to social interactions deals in morality. A philosophical industry has been built around vexing moral quandaries such as the well-known trolley problems. In one version, a train loses its brakes while approaching a fork in the track. You operate a switch that directs a car onto one side or other of the split. On the right side there are five people on the track whom the trolley will fatally strike, were it to go that way. If the car goes left, it will deliver a deadly blow to one person. What do you decide to do with the switch? Is there a right choice? Is it even a moral issue? Is this a difficult decision? Why or why not? From our point of view, does symmetry play into the situation? If so, how?

By ruminating on the nature of morality, Immanuel Kant worked out a theory for this thorny subject [27]. He built such an account on a foundation of two principles. The first is a statement of value: humans—perhaps, all sentient beings—possess intrinsic worth and, as such, should never be treated as a means to an end. The second pillar of moral philosophy gives us a tool that can tell us whether an action is one to which we're duty-bound to perform. Here, the key idea is that a morally correct act is *universalizable*—an awkward term that relies on population effects. To characterize the concept, we refer to an agent P's moral community, which is the network of agents who experience some effect from an action taken by P.

> Action A performed by P is universalizable when *every* member of P's moral community can approve of P's performing A.

The subjunctive 'can' is worth noting. According to Kant, universalizability (whew!) confers a *categorical imperative* on a moral agent. That is to say, someone should act provided that everyone can agree to it. If we apply the principle to the trolley problem, neither option is universalizable. Switch right and the five people on the track don't approve. Switching left does not meet with the approval of the person on that side. It looks like Kant would say that throwing the switch is not a moral choice.

Kant's treatment of ethics has been the subject of extensive study and commentary. Our modest goal is to locate his account in a choose-first setting. Take an agent's action to be a choice that figuratively passes to their community in order to calculate its moral rectitude.

Drawing on Kant's notion of a population-wide process, consider another basic ethical principle that we call a *universal hypothetical*.

> An action A by P is morally right provided that no one suffers harm were *every* member of P's moral community to perform A.

Here, again, treat an action to be morally evaluated as a choice. The new rule would have all relevant agents engage in a kind of simulation the output of which yields the action's moral status. A few sample cases might be helpful. Following traffic regulations is easily worked out. Is it wrong to abstain from voting in a democracy? If every citizen did so, democratic structures would be impossible. When visiting a popular mountain peak, is it alright to take away some of the rocks that are strewn about? What would the place look like if every visitor did the same? Compare this scenario to removing grains of sand from a beach. Does the universal hypothetical seem to work out as it does on the mountain?

The hypothetical reasoning exhibited here is fundamental to science. Given a set of assumptions, work out their logical consequences. This process is akin to employing a model to simulate a system's behavior.

8.3 GIVE CHOICE A CHANCE

Our prior discussion of evolutionary processes regarded natural selection as a kind of computational device that leads to a chosen phenotypic outcome. Here, we reverse the view by casting evolution as an algorithmic system that operates on an entity that we choose at the outset. Since we develop a procedure that closely tracks the dynamical method of equation-solving, our treatment of evolutionary algorithms goes into much finer detail than other topics. In particular, we use some formal notation and work through counting issues that are a bit technical.

A robot's problem

Autonomous machines that sweep floors are a recent technology that has achieved considerable popularity. At any given point in the device's operation, the basic issue is deciding where to go next. The hard problem is to work out a set of rules that guide the sweeper to near optimal performance.

"Organism"

We imagine a somewhat idealized environment consisting of a hexagonal grid of appropriate sized cells on which a robot called "hexabot" operates. Each cell is in one of two states: clear or blocked. The robot can occupy only clear cells. Given a cell C, a *scene* is an assignment of states to the six cells adjacent to C. Figure 8.2 displays several scenes where blocking obstacles appear as solid dots. Note the cell labels 1 to 6 running counter-clockwise around the ring of cells. Using 0 and 1 for blocked and clear, describe a scene as a sequence of six binary digits (bits). Below each scene diagram in the figure, two rows of digits depict this formulation, from which it follows that the total number of scenes is $64 = 2^6$. The row on top records the states of cells labeled according to entries in the bottom row. For convenience, the labeling places 1 in the cell toward which hexabot's "nose" points. Take the sample at upper left. Since cells 1 and 2 are blocked while 3 through 6 are clear, the binary expression for this condition—reading along the top row from right to left—places 0 in positions 1 and 2 with 1 at each of the other four. Observe that the scene expressed by 000000 with all cells blocked is not worth considering since hexabot has no where to go. Therefore, we work with the remaining 63 scenes.

Hexabot moves by "choosing" one of the clear cells among the six adjacent to the one it currently occupies. Divide a move into two parts: first rotate so that hexabot points toward the cell prescribed by its instructions and then cross into that cell while maintaining direction in order to establish cell labels. Figure 8.3 illustrates such a motion.

Next, we create a set of instructions called an *evolutionary protocol* (or just "protocol") that guides hexabot's movement for any scene that it encounters. Doing this requires us to prescribe an appropriate action for every scene. As such, you can construe a protocol as a function from the set of 63 scenes to the set of actions. There is, however, a more computation-friendly description.

Say that a move is *permitted* when it avoids all neighboring cells that are blocked. We allow a protocol to dictate motion onto any clear cell or to randomly select one of the clear cells. Accordingly, the number of actions associated with a scene is given by the number of clear cells plus one for a

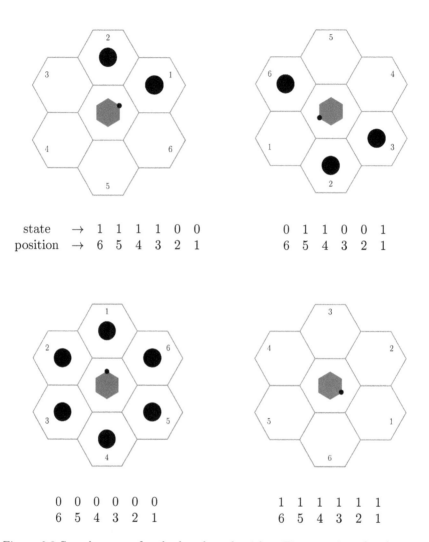

state   →   1 1 1 1 0 0   0 1 1 0 0 1
position →   6 5 4 3 2 1   6 5 4 3 2 1

0 0 0 0 0 0   1 1 1 1 1 1
6 5 4 3 2 1   6 5 4 3 2 1

Figure 8.2 Sample scenes for the hexabot algorithm. Binary strings for the scenes appear below a diagram with a 0 (blocked) or 1 (clear) above each position label. Hexabot's nose always points toward cell-1. We exclude the scene given by 000000 since it traps the robot.

random move. Being analogous to an organism's genome, protocols are central to an evolutionary algorithm. Some notation can help spell out how to work with them. Using their binary descriptions, scenes have a natural ordering from 0 to 63. List all scenes in order and assign an action to each as Table 8.4 indicates.

Figure 8.3 Two-step hexabot motion for protocol instruction $100111 \to M3$. The action here is a move to cell-3.

To gain a better sense of protocol behavior, let's see how hexabot can get stuck on a pair of cells. A simple way this can happen is for two adjacent cells to be surrounded by obstacles. Since hexabot can neither escape nor enter this region, we prohibit such an enclosure as well as arrangements of obstacles that surround other clusters of cells. In other words, a "path" along adjacent blocked cells cannot form a loop. Figure 8.4 illustrates a non-trivial instance of a period-two trap. Relevant lines in an active protocol prescribe actions leading to repetition in the second and third moves.

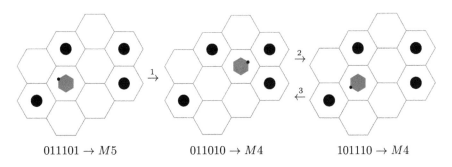

$$011101 \to M5 \qquad 011010 \to M4 \qquad 101110 \to M4$$

Figure 8.4 Hexabot can get stuck. Labels on arrows correspond to the order in which moves occur. Note how the protocol's binary string instruction depends on hexabot's orientation.

For a challenge, the reader can attempt to design a protocol that runs on a grid that contains no obstacles. How many scenes can there be? Does it matter where hexabot starts?

Population

By arranging scenes in a particular way, we can now count the number of protocols. Put scenes into groups according to how many clear cells each one carries. Quantifying the number in such a group agrees with our counting of particle configurations in Section 2.11. For instance, suppose that a scene

Table 8.4 Creating a Protocol. Scenes are ordered according to their binary representation with the base-10 equivalent. Arrows stand for the assignment of actions to scenes. The symbols $M1$, $M2,\ldots,Mk$ mean that hexabot moves to cell 1, 2,...,k. Use $R(c_1, c_2 \ldots, c_k)$ to indicate a random choice of clear cell c_1, c_2, ..., or c_k.

Number	Scene		Possible actions (choose one)
0	000000	$\longrightarrow$	disregard
1	000001	$\longrightarrow$	$M1$
2	000010	$\longrightarrow$	$M2$
3	000011	$\longrightarrow$	$M1, M2, R(1, 2)$
4	000100	$\longrightarrow$	$M3$
5	000101	$\longrightarrow$	$M1, M3, R(1, 3)$
⋮	⋮	⋮	⋮
29	011101	$\longrightarrow$	$M1, M3, M4, M5, R(1, 3, 4, 5)$
30	011110	$\longrightarrow$	$M2, M3, M4, M5, R(2, 3, 4, 5)$
31	011111	$\longrightarrow$	$M1, M2, M3, M4, M5, R(1, 2, 3, 4, 5)$
32	100000	$\longrightarrow$	$M6$
33	100001	$\longrightarrow$	$M1, M6, R(1, 6)$
⋮	⋮	⋮	⋮
59	111011	$\longrightarrow$	$M1, M2, M4, M5, M6, R(1, 2, 4, 5, 6)$
60	111100	$\longrightarrow$	$M3, M4, M5, M6, R(3, 4, 5, 6)$
61	111101	$\longrightarrow$	$M1, M3, M4, M5, M6, R(1, 3, 4, 5, 6)$
62	111110	$\longrightarrow$	$M2, M3, M4, M5, M6, R(2, 3, 4, 5, 6)$
63	111111	$\longrightarrow$	$M1, M2, M3, M4, M5, M6, R(1, 2, 3, 4, 5, 6)$

contains three clear cells so that its binary expression has three 1s to which we refer as first, second, and third. There are six places where the first 1 can go, for each of which the second has five possible locations. These 30 arrangements of the first two digits leave four spots for the third 1, resulting in $120 = 6 \cdot 5 \cdot 4$ configurations. Since we aren't really distinguishing between the digits, we need to divide by the number of ways that three objects can undergo shuffling, that is, $6 = 3 \cdot 2$. Thus, we can place three 1s into a six-digit binary sequence in

$$\frac{6 \cdot 5 \cdot 4}{3 \cdot 2} = 20$$

different ways. Table 8.5 presents the remaining cases.

To work out the quantity of protocols, take note of how many actions there are for each type of scene. As an illustration of the counting process, consider the situation when there are two clear cells. For each of the fifteen scenes with that many openings, three actions can occur, namely, move to either of the free cells or to one of those cells chosen at random. Therefore, there are 3^{15} ways of assigning permitted actions to these fifteen scenes. The table refers

Table 8.5 Counting Protocols.

# Clear cells	Scenes	# Scenes	# Actions	Protocol factor
1	000001 $\vdots$ 100000	6	1	1^6
2	000011 $\vdots$ 110000	$\dfrac{6 \cdot 5}{2} = 15$	3	3^{15}
3	000111 $\vdots$ 111000	$\dfrac{6 \cdot 5 \cdot 4}{3 \cdot 2} = 20$	4	4^{20}
4	001111 $\vdots$ 111100	$\dfrac{6 \cdot 5 \cdot 4 \cdot 3}{4 \cdot 3 \cdot 2} = 15$	5	5^{15}
5	011111 $\vdots$ 111110	$\dfrac{6 \cdot 5 \cdot 4 \cdot 3 \cdot 2}{5 \cdot 4 \cdot 3 \cdot 2} = 6$	6	6^6
6	111111	1	7	7^1

to this value as a protocol factor, since the total number of protocols is found by multiplying all of these factors together. For a result, we get about 10^{35} distinct programs that can govern hexabot's behavior.

A population of protocols is the entity that evolves when our algorithm runs. However, since working with such a large "protocol space" is intractable, we're compelled to work with a much smaller collection. This restriction is a weakness in evolutionary computation, one that sampling practice can offset to some extent.

Fitness

Before describing how an algorithm confers development on an aggregation of protocols, we need to spell out the evolutionary goal that it pursues. (Such teleology is not a feature of biological evolution.) The procedure's design is to increase protocol *fitness*, a property that it evaluates with a score. As indicated in Figure 8.5, place a programmed hexabot on an obstacle-laden hexagonal grid that represents a room. Treat the cell edges that are not between two

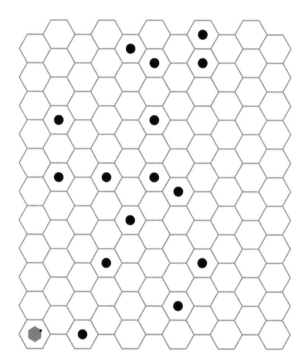

Figure 8.5 Space in which the hexabot acts.

cells—that is, the ones that form the grid's boundary—as blocked. Beginning at the initial position, allow the protocol to run until hexabot's itinerary meets one of three conditions.

1. Transitivity: hexabot visits every clear cell.[1]

2. Periodicity: hexabot returns to a cell in the same orientation.

3. Time-out: hexabot executes a specified number of moves.

Use a two-coordinate description of the grid cells to record the resulting path

---

[1]This refers to a dynamical notion of transitivity.

and assign it a fitness score according to a rubric such as the one found in Table 8.6. The idea is to penalize an itinerary for either missing or revisiting a spot. We can promote robustness in a population by running each protocol on several initial positions for a variety of grid configurations, each with the same number of clear cells. Afterward, you can use more or less clear cells. For an overall fitness score, average the achieved values. There are 106 clear cells in the featured grid so that the maximum score is $1590 = 106 \cdot 15$. Note that this is a theoretical maximum since it might not be possible to visit every cell exactly once.

Table 8.6 Scoring Protocol Fitness.

# Visits to a given cell	Points
0	-5
1	15
2	$14 = 15 - 1$
3	$12 = 15 - 1 - 2$
4	$9 = 15 - 1 - 2 - 3$
5	$5 = 15 - 1 - 2 - 3 - 4$
6+	$0 = 15 - 1 - 2 - 3 - 4 - 5$

Selection under variation

An essential factor in an evolutionary system is a means by which a population undergoes heritable variation. Our computational "organisms" or "genomes" are subject to splicing and mutation, forms of alteration into which inheritance is built. Begin with a random collection of, say, 100 protocols. This number is selected for convenience. Perhaps 1000 or 10, 000 is feasible, but these are still relatively tiny samples compared to the whole protocol space. Using a larger population incurs a computational cost, but makes it more likely that protocols with high fitness will arise. Now that we have a method of modification from one generation to the next, it's time to spell out how to run an evolutionary algorithm.

Take several allowable grid configurations with equal numbers of clear cells. Using the same set of initial conditions—cell location and hexabot orientation, score each protocol's itinerary and let the average be a fitness score. Now, rank the 50 top-scoring protocols and list them in descending order:

$$P_0 = \{p_{0,1}, p_{0,2}, \ldots, p_{0,49}, p_{0,50}\}.$$

Use of a double index suggests that the notation here can become rather baroque. Let's resist the temptation to go that way. Think of P_0 as the zeroth or initial generation that, by modification, gives rise to a first generation P_1. The left index 0 attached to the listed protocols provides generational information and the right one gives a protocol's rank. So equipped, we can engineer the reproductive process.

Create a new population of 100 protocols by splicing together fragments of the members in P_0. There are many ways to graft the pieces together. To illustrate a simple version of how it can be done, take the top two protocols

$$p_{0,1} = (a_1, a_2, \ldots, a_{29}, a_{30}, \ldots, a_{62}, a_{63})$$
$$p_{0,2} = (b_1, b_2, \ldots, b_{29}, b_{30}, \ldots, b_{62}, b_{63}).$$

The symbols a_1 to a_{63} and b_1 to b_{63} stand for actions assigned to scenes numbered 1 to 63 in Table 8.4. Cut each protocol at some location, $29|30$, say, and build two new sequences by swapping "tail" fragments:

$$(a_1, a_2, \ldots, a_{29}, b_{30}, \ldots, b_{62}, b_{63})$$
$$(b_1, b_2, \ldots, b_{29}, a_{30}, \ldots, a_{62}, a_{63}).$$

Notice that an action permitted for a scene in one protocol is also allowed for the same scene in the other protocol. For example, a_{30} and b_{30} are both valid actions for scene 30. Repeating the process with $p_{0,1}$ and $p_{0,2}$ at another cut-point generates two more protocols. Continued cutting and splicing on

$$p_{0,3} \text{ and } p_{0,4} \qquad p_{0,5} \text{ and } p_{0,6} \quad \cdots \quad p_{0,49} \text{ and } p_{0,50}$$

produces four new protocols for each of the 25 pairs, yielding a novel set of 100 protocols. Before putting them through their paces, allow random mutation to occur.

Pick a mutation rate m between 0 and 1, typically close to 0. Consider each scene in a new protocol and with probability m, replace the originally prescribed action with a permitted one that's obtained at random. If $m = .02$, for example, a protocol would be expected to admit roughly one mutant site. Suppose $a_{29} = M_4$ and that chance calls for a mutation at this point in the respective protocol genome with M_2 as replacement. The alteration looks like this:

$$(a_1, a_2, \ldots, a_{28}, \quad M_4 \quad , b_{30}, \ldots, b_{62}, b_{63})$$
$$\downarrow$$
$$(a_1, a_2, \ldots, a_{28}, \quad M_2 \quad , b_{30}, \ldots, b_{62}, b_{63}).$$

We can apply the same procedure to the newly spliced set of 100 protocols. Running, scoring, and modifying them leads to a first generation population

$$P_1 = \{p_{1,1}, p_{1,2}, \ldots, p_{1,49}, p_{1,50}\}$$

ranked by fitness. Recursively processing *run-score-modify* manufactures successive generations of protocol clusters P_2, P_3, and so on. Should we expect to see the emergence of protocols with greater fitness? At the beginning, randomly sampled members of P_0 are likely to have low fitness, since a low score seems to be more easily achieved than one that's high. After modification, there are more ways for a protocol's performance to improve than decline. Accordingly, P_1's most fit protocol (MFP) is probably more fit than P_0's MFP. This amelioration ought to persist while the MFPs have relatively low fitness. But, as their scores increase, it becomes more difficult to generate protocols with still greater fitness. Eventually, there are more ways that fitness can decline than grow. We anticipate that successive MFPs will eventually have fitness scores that tend to a plateau. Whether the value associated with this flattening is near the maximum score looks like a difficult theoretical and computational question. Were the score close to optimal, variation in the population would tend to diminish fitness by a relatively small amount. Another puzzle concerns the associated sequence of MFPs; does it reach a point in protocol space for which nearby points are less fit? Such a location is known as a local maximum, but it need not be globally so.

In any case, the tendency of MFP fitness toward a particular level exhibits attracting dynamical behavior. A comparison between protocol evolution and equation-solving dynamics is striking as both exemplify choose-first algorithms. You can view run-score-modify as a map that takes a random initial condition P_0 and generates a dynamical orbit in protocol space that, hypothetically, finds a point with high fitness. If such a protocol is optimal, it should look like an attracting fixed point.

By simply inspecting every protocol's set of instructions—itself an unfeasible task, you would find it difficult to determine their fitness levels. Similarly, just looking at points in a space associated with a polynomial's symmetries, it's impossible to tell which ones can be used to harvest roots. The number of choices is overwhelming, somewhat like deciding among many comparable types of cereal. On this point, having a lot of options can signify a type of statistical symmetry whereby the items can seem to be indistinguishable due simply to the sheer number of them. Call it a symmetry of large numbers.

In both polynomial and protocol contexts, the only remedy to the large-number impediment might be to let dynamics take its course on points in their respective spaces. There is, of course, a significant difference between the two settings, namely, the number of points in the space of protocols is finite.

What might we hope to derive from implementing a machine learning program based on evolutionary principles? Expecting to find a protocol that reliably achieves near maximum scores is probably optimism in excess. We can anticipate the appearance of some protocols whose scores tend to a fairly high level, though we aren't assured of this outcome. Since a protocol is specified by 63 values, as Table 8.4 indicates, the dimension of the space in which it lies is also 63. Attaching a fitness score to each point (protocol) in this space adds a dimension, and so constructs a 64-dimensional "protocol landscape."

Following an evolutionary path will, we hope, move an evolving population's MFP uphill in order to reach the highest peak. However, the landscape could be such that the trajectory of successive MFPs could get stuck on a "local peak" where nearby points have lower scores, but a distant protocol stands at a higher elevation. Some adjustments might nudge a stuck protocol orbit far enough so that it finds a way to enhanced fitness. Running the algorithm for more generations is the simplest thing to try. Alternatively, we could slightly perturb the mutation probability or the splicing location.

Another tactic involves searching the highest-scoring protocols for behavior patterns that suggest ways to change a few rules. Consider a couple illustrations of such patterns. (Disclaimer: these principles are offered solely for purposes of demonstration. There is no claim that a high-scoring protocol conforms to them.)

1. When a scene contains a clear cell adjacent to two that are blocked, don't move there, if possible.

2. If three consecutive clear cells appear in a scene, move to the middle cell of the three. If more than one triple of adjacent clear cells exists, move randomly to one of the middle cells.

If some of the rules followed by a high-scoring protocol agree with a principle such as these, try aligning a small number of the other rules accordingly.

Problem in a crowded bar

To introduce another circumstance on which an evolutionary algorithm might provide illumination, consider two friends A and B who arrange to meet at a bar that turns out to be rather busy. After both arrive, they find themselves on opposite sides of the room. Being mathematicians, their goal is to reach one another using a procedure to navigate through the crowd in an optimal way. They realize that more than one such algorithm might exist. Figure 8.6 depicts the environment as a hexagonal grid with other patrons acting as obstacles denoted by solid dots.

The first issue to resolve is how to decide when one method is better than another. An obvious standard is the number of moves it takes for the parties to meet going from cell to cell. A grid's configuration dictates a lower bound to this number. For the layout shown in the figure, they can be on adjacent cells in six steps and no fewer. Paths that minimize the number of steps appear in Figure 8.7; note that such an optimal route is not unique. They want to work out a set of rules for moving from one cell to the next that brings them together in close to a minimum number of steps. From a bird's eye view, you might readily see how they can do this. A harder problem is for each of them to decide how to move when all they can see are the states of neighboring cells—clear or blocked—and what direction points toward their friend.

They opt for an evolutionary approach. To set things up, we need to describe scenes and protocols, ingredients that work as in the hexabot algorithm but with some nuance. In the present case, a scene consists of a hexagonal hub with its six adjacent cells plus information concerning the direction in which the target cell lies. To make the description simpler, distance to the target is not provided, although it could be. Figure 8.8 shows scenes for A and B that

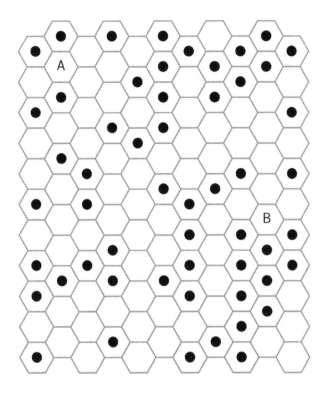

Figure 8.6 In a crowded bar, how can friends A and B meet by following an efficient algorithm?

correspond to the configuration expressed in the previous diagram. Arrows indicate where one friend looks to see the other. Labels for the six "satellite" cells are all oriented the same way relative to the room. From right to left, the first six digits in the scene description conveys the same information as the one used for hexabot. The value at position 7 tells us to which edge of the hub the arrow is pointing. Counting the number of scenes proceeds as before; six binary slots again give 63 and for each of these there are six directional values, yielding a total of $378 = 63 \cdot 6$. By a suitable adaptation mentioned below, we can include the scene in which all cells are blocked.

As for constructing protocols, we augment the set of hexabot actions to include remaining in place either as a deterministic or random event. Table 8.7 is comparable to a previous one and displays the effect of this embellishment. Prescribing such a null action in the floor-sweeping system would fix hexabot's position for the remainder of time. In the bar setting, a move by B can alter A's scene even though A was instructed not to change location. Having two parties exert mutual influence makes for considerably more complexity than before, as we see next by calculating the size of protocol space.

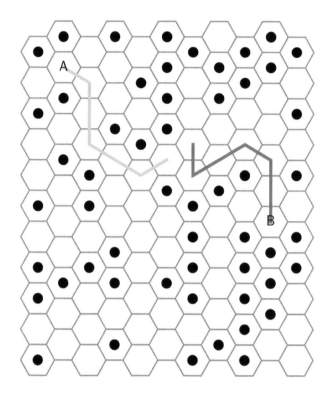

Figure 8.7 Shortest paths for friends to meet—to be on adjacent cells after a given number of moves.

A scene's binary piece (positions 1 to 6) determines which moves are permitted. Holding constant the states attached to those positions, there are six possibilities that can occur at position 7. If you pair the various types of scene with the corresponding number of actions—as worked out previously in Table 8.5—and then multiply the numerical results, you end up with approximately 10^{261} protocols. This space contains vastly more points than the one for hexabot protocols and, once again, is far too large for us to "happen upon" an efficient protocol. Rather, we look for one to emerge from an evolving population that's dynamically driven by the same recursive steps taken by the previous algorithm. Each new variant generation undergoes fitness evaluation in which taking fewer steps counts as a better score, provided that a protocol succeeds in bringing A and B together. The most fit protocols are selected to pass their genomes onto the following generation. If the sequence of MFPs actually reach relatively high fitness, we can inspect them for behavioral patterns that suggest ways in which they might be favorably tweaked. Also worth seeking is a general principle such as: Always move toward your friend to the greatest degree possible. How would you code for this behavior in a protocol?

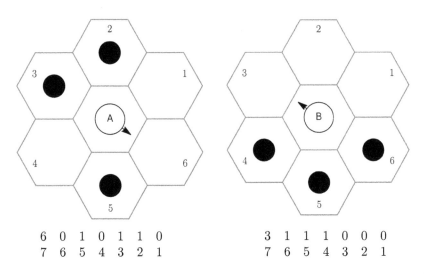

6 0 1 0 1 1 0
7 6 5 4 3 2 1

3 1 1 1 0 0 0
7 6 5 4 3 2 1

Figure 8.8 Bar scenes. These configurations agree with what Figure 8.6 shows. Each friend's arrow points at the other.

Table 8.7 Creating Another Protocol. In the "scene" column, the entry denoted (1–6) abbreviates listing six strings whose leftmost or seventh position is a digit between 1 and 6. The column of "possible actions" is also amended to reflect the remain-in-place option designated by $M0$. A null move is also one of the acceptable random outcomes.

Number	Scene		Possible actions (choose one)
1–6	(1–6)000001	$\longrightarrow$	$M0, M1, R(0,1)$
7–12	(1–6)000010	$\longrightarrow$	$M0, M2, R(0,2)$
13–18	(1–6)000011	$\longrightarrow$	$M0, M1, M2, R(0,1,2)$
19–24	(1–6)000100	$\longrightarrow$	$M0, M3, R(0,3)$
25–30	(1–6)000101	$\longrightarrow$	$M0, M1, M3, R(0,1,3)$
⋮	⋮	⋮	⋮
349–354	(1–6)111011	$\longrightarrow$	$M0, M1, M2, M4, M5, M6,$ $R(0,1,2,4,5,6)$
355–360	(1–6)111100	$\longrightarrow$	$M0, M3, M4, M5, M6,$ $R(0,3,4,5,6)$
361–366	(1–6)111101	$\longrightarrow$	$M0, M1, M3, M4, M5, M6,$ $R(0,1,3,4,5,6)$
367–372	(1–6)111110	$\longrightarrow$	$M0, M2, M3, M4, M5, M6,$ $R(0,2,3,4,5,6)$
373–378	(1–6)111111	$\longrightarrow$	$M0, M1, M2, M3, M4, M5, M6,$ $R(0,1,2,3,4,5,6)$

Another consideration is whether the protocol populations for each party should evolve independently. Computationally, having them follow the same rules is less expensive. However, if they use sets of protocols that develop separately, there might be a greater opportunity for high fitness to emerge.

The method we just described suffers from a shortcoming, namely, that positions of the bar patrons other than A and B are static. Under such a restriction, the friend-finding problem has the flavor of percolation—the study of how fluid flows around an arrangement of fixed obstructions. In a more realistic model, the other folks in the bar would move about to some extent. However, creating a set of rules that authentically governs their movement is not a simple matter. Indeed, evolutionary dynamics could be brought to bear on this problem within a problem. An approach that lacks realism has the obstacles move to an adjacent cell with some probability. Such a dynamical model allows the friends to benefit from remaining in place, either as a deterministic or random action.

A significant difference between floor sweeping and crowded bar algorithms pertains to the means by which hexabot and either friend gather information from their respective scenes. For the former, only local data are available, whereas the latter allows for communication from both local and limited distant sources. In these contexts, two cells are "local" relative to one another if they share an edge. The dynamical impact of non-local interactions on evolving protocol phenotypes might make a fruitful exploration. To see what evolution can invent, the reader is invited to create computer code that implements these algorithms.

Evolution computes

Evolutionary computation provides a rich agent-based ecosystem in which to embed and explore certain optimization problems. The agents in such a complex system are the entities that evolve, namely, collections of protocols. You may not be likely to use an evolution-driven program in order to choose between restaurants or brands of cereal, but they can give you a means to finding a somewhat fit set of rules in a space that contains a bewildering number of such sets. Many problems whose complexity and level of difficulty places them beyond the reach of intuition and pure analysis might be amenable to a learning algorithm such as one that relies on evolutionary mechanisms. To suggest a few:

- How do proteins that are manufactured in a cell interact and organize themselves into function-performing structures? Perhaps an evolutionary algorithm can illuminate actual biological evolution.

- How should an autonomous stock cart navigate through a warehouse in order to load a selection of goods efficiently?

- Devise a set of rules (locally-based or not) that guides members of a swarming population such as flocking birds or schooling fish. Working out lifelike movements of patrons in a crowded bar fits into this category of problem.

Going at one of these challenges with an algorithm that harnesses evolutionary development requires a series of steps. First, find a suitable environment—either concrete or abstract—on which to model the problem. Characterizing what constitutes a scene and an action is next. From there, build a protocol space. Given a notion of fitness—typically, a quantitative property that the process optimizes, recursively subject populations of protocols to determinate as well as random variation and then to selective pressure. Finally, watch for the emergence of highly fit protocols.

What about a problem for which an evolutionary approach founders? Consider a chess-playing machine for which the board itself is a natural environment. The only reasonable candidate for a scene is the arrangement of pieces across the entire board, which makes the set of scenes so large as to be unworkable when it comes to constructing protocols. Perhaps evolution would come up with something if the pieces are sparse on the board.

CHAPTER 9

Conclusion: The Price We Pay for Symmetry

For the most part, our examination of choice follows a theoretical course. Some of the discussion has pragmatic value. In certain situations, moral algebra can be useful, especially when symmetry is absent. Concrete measures known as nudges can significantly influence decision-making [43]. Far more subtle, sophisticated, and powerful are evolutionary algorithms.

This work's primary goal is to illuminate the role that choosing can play when we take on challenging problems. To conclude the exposition, we entertain some big-picture thoughts that tie together its principal themes: namely, choices belong to spaces, groups of symmetries can act on such spaces, and choose-first methods dynamically refine symmetry-breaking initial choices.

9.1 SYMMETRY, MORE OR LESS

We introduced the seminal concept of symmetry in order to characterize polynomials according to their structures. The algebraic tool used for this purpose is the symmetry group, expressed as ways of shuffling an equation's solutions. Through the medium of their symmetry groups, we then associate polynomials with geometric objects—mostly polyhedral configurations on a sphere. The final step of this choose-first approach constructs maps whose dynamical symmetry matches that of the equations they are instrumental in solving.

The symmetries considered here are exact in the sense that they transform an infinite space of choices so that one or more of a resident object's properties are preserved perfectly. For instance, under the action of a rotation, the anatomical elements of a polyhedron as well as its overall position remain precisely the same. Such is not the case in choose-first circumstances like those imagined in Part II. By way of illustration, take one of our evolutionary algorithms. The space in which choices reside consists of points that represent protocols. While it's surely easy to tell that some sets of rules have poor fitness, a large number—likely, very large—possess fitness credentials that are mutually difficult to differentiate. Denote such a cluster of points as X. Here's

DOI: 10.1201/9781003098164-9

a way to think about symmetry relative to a finite but immense protocol space. It didn't come up in our treatment, but protocols—like scenes—can be ordered numerically. The details aren't important. From the set X, extract the first 100 points as an initial condition for evolutionary processing. Next, shuffle X's elements in any way and then draw out the first 100 protocols. Our inability to distinguish members of X on the basis of fitness means that we should expect the two samples to evolve in comparable ways. The source of this shuffle symmetry is found in the complexity of the space on which evolutionary dynamics develops.

One place where procedures that solve equations differ from those employed in a context beyond equations has to do with dynamical symmetry. There's a precise statement of what it means for a map to have symmetry of some kind. We saw that a dynamical system at the core of an equation-solving process has an explicit form, the symmetry of which satisfies a specific algebraic condition. As a consequence, such a map generates dynamical orbits that are deterministic, without a trace of random behavior. Its definite symmetrical action on a relevant space is responsible for the basins-of-attraction graphics with which we are rewarded in Chapter 5.

In contrast, this sort of symmetry is not shared by recursive choose-first procedures such as playing strategy-sets, running simulations on configurations of congressional districts, or urging collections of protocols to evolve greater fitness. These methods invoke maps with features that are stochastic in nature and crucial to the function that such methods perform. Lacking deterministic behavior, each of these maps fails to admit a symbolic description expressed in pure algebra and so, its symmetrical status is not easily characterized. Nevertheless, there might be a probabilistic theory that articulates a version of *stochastic symmetry* suitable for the task at hand. Working out such an account looks like a tough nut to crack.

9.2 CHOOSING AS METAPHOR

An elementary world

When thinking or talking about making a choice, we have a sense that it's a discrete act. We come to a fork in the road and then select one of the alternatives to take. Clearly, choosing has a digital quality, like throwing a light switch. Viewed this way, the course of someone's life could be mapped out as a path across a tree where each branch that's followed corresponds to a choice made. But, maybe the act of choosing is more of an analog process—where a dart lands on a board, sliding a dimmer switch and not quite being sure of when a light is on. To make a selection, you take into account many paths in a space of choices that looks like a continuum where you can take arbitrarily small steps. Following the trajectory of a life still involves branching that the act of choosing creates. But, until a clear branch appears, all possible options are explored.

In quantum mechanics, such exploration is realized as a superposition of an object's states. Branching occurs when a quantum system undergoes measurement. Consider a beam of light that strikes a pane of glass. If you view what happens from a certain location on the side that contains the light source, a reflected beam appears. From the pane's other side, you see a beam that passes through. The same phenomenon occurs when you look at a window and perceive both your reflection and what's on the other side of the glass. Light from one side both reflects from and transmits through the glass. Treating light as propagating electromagnetic fields, this empirical result can be explained by considering the interaction between an oscillating electric field and the matter found in the glass. However, quantum electrodynamics regards a light beam as a stream of particles called photons. What fraction of the incident photons is reflected back? Or put more carefully, how much light scatters due to its interaction with matter in the glass? Regardless of how many photons there are, the result is always the same value—maybe around 4%. So then, what happens to a *single* photon just as it reaches the glass surface? How does it "decide" whether it belongs to the 4% that scatters back or the 96% that scatters through? The best answer currently on offer—given in the section that follows—shouldn't surprise a careful reader. According to theory, the particle follows all possible paths from source to target. Each route is described by a complex number called an amplitude, which is a quantity that captures the probability that a quantum of light follows a particular path. There are phenomena that we can measure such as a photon's straight-line trajectory that either goes through the glass or bounces back. Outcomes of this sort result from a process of adding up the amplitudes of all paths that the photon takes. In a sense, the observed phenomenon arises from summing all routes that a photon can choose [20].

Another fundamental choice that confronts nature concerns whether objects with mass are composed of matter or antimatter. What makes the decision a perplexing one is a mirror-like symmetry. After undergoing a transformation that reverses positive and negative charge rather than left and right orientation, a particle of antimatter is a bit like a reflected version of a matter particle. As such, the basic problem is a cosmic variation on a Buridanic theme. According to the matter-antimatter symmetry, the two should have been equally plentiful and distributed in the early universe. However, these conditions would have led to the rapid annihilation of material substance as well as everything that depends on it, including stars and planets. When a particle of matter collides with one of its antimatter partners, the pair disintegrates into photonic debris. If the universe is not to be starved of mass and the mule is not to starve, symmetry-breaking choices must occur in a spontaneous way.

Life's choice

In anthropomorphic terms, evolution can seem to be clever at solving problems. An explanation for this "intelligence" is found in the way that evolutionary processes try many genomic paths, like and unlike a quantum particle that explores *all* physical routes between states. Every mating and mutation that a population experiences is an experimental choice, one that undergoes (computational) testing at the hands of an ecosystem. Environmental conditions select for traits by acting on what results from evolutionary choices, whose source is discussed below. You might say that evolution is doubly selective.

An essential product of human evolutionary history is a highly complex system of immunity. Recognizing and counteracting an infectious substance is, of course, the central problem facing an immune response. One of the crucial steps taken to meet this challenge is the creation of antibody factories known as plasma B cells. Each of these white blood cells is tasked with manufacturing one type of protein that can latch onto a specific antigen—a piece of protein on a particular virus, say. There are a staggering number of microbes with which we can be infected so that B cells need to make a comparable amount of antibodies that can protect against *any* particular microbial incursion. To accomplish the job, they use a technique that relies on combinatorial fecundity, which is a property seen previously.

B cells employ their methodology at the level of genes, effectively generating genetic novelty by "choosing" from a menu of DNA fragments. By analogy, imagine that you want to make a lunch consisting of a sandwich and fruit. You have five kinds of bread and three types of spread for making sandwiches. There are seven fruits from which to choose. How many distinct lunches can you make? From earlier discussions of similar questions, the number is $5 \cdot 3 \cdot 7 = 105$. Let each of the three ingredients correspond to a category of DNA fragment. A gene that codes for an antibody appears when the B cell prepares a lunch by selecting and joining together one segment from each category. In actuality, there are more categories of fragments as well as more pieces of DNA in each category. Thus, there are many more genes—hence, antibody proteins—that the cell can construct. In the final section, we address the way in which a B cell chooses.

9.3   RANDOM CHOICE IS UNAVOIDABLE

If our exploration into equation-solving and beyond has taught us something, it is that certain problems are difficult because they call for making hard choices. Furthermore, some kind of symmetry is often responsible for the trouble, by setting up an obstruction that must be cleared.

From Buridan's mule to the scattering of light to evolutionary innovation, a core lesson from solving polynomial equations can be applied to these elementary problems. In every one of these cases, and many more besides, the role of random choice is crucial to solving difficult problems.

When a photon strikes a window pane, it randomly chooses to scatter either toward or away from the side on which it appeared. It's as though the particle rolls a 100-sided die—analogous to summing amplitudes—and opts for reflection if a side 1 to 4 turns up; otherwise, transmission through the glass takes place.

As for the matter-antimatter puzzle, it must be that the universe exercised in random fashion a slight bias in favor of matter. The cosmological nudge that broke symmetry remains mysterious.

Biological reproduction is inherently influenced by random events. Which two organisms mate is somewhat haphazard, albeit not entirely the result of random interactions. Mutation counts as a more thoroughly accidental occurrence, such as an exposure to high-energy radiation or a seemingly spontaneous copying error. One of Darwin's great insights is that evolution need not have a direction. Rather, moving incrementally and randomly in an enormous space of genomic possibilities suffices for a great variety of species to evolve. Perhaps this movement is only partially incremental and random, as there can be significant phenotypic jumps and mutations that are more likely to occur in light of previous modifications.

To build an arsenal of antibody proteins, a B cell selects a random assortment of gene fragments for each of their respective types. Antibody space is huge and possesses high entropy since any one combination is statistically equivalent to any other. Once again, choosing randomly provides a viable mechanism that overcomes the symmetry that underlies this equivalence.

From Galois we learn of a polynomial's symmetry and how it exacts a cost when solving the corresponding equation. Following symmetry's extensive conceptual thread leads us to a geometric understanding of polynomials. On that foundation sits a dynamical edifice that solves this classic problem by harnessing the symmetry-breaking nature of a random guess. Stepping outside the realm of equations, we find consonance between structures in mathematics and decision-making. Perhaps the ultimate price we pay for symmetry is the acceptance that many of our momentous choices can be made only by the intervention of chance.

Bibliography

[1] M. Armstrong. *Groups and Symmetry*. Springer-Verlag, New York, 1988.

[2] A. Beardon. *Iteration of Rational Functions*. Springer-Verlag, New York, 1991.

[3] C. Borgers. *Mathematics of Social Choice: Voting, Compensation, and Division*. Society for Industrial and Applied Mathematics, Philadelphia, 2010.

[4] H. Burgeil, J. Conway, and C. Goodman-Strauss. *The Symmetries of Things*. CRC Press, Boca Raton, 2008.

[5] G. Cardano. *The Rules of Algebra*. MIT Press, Cambridge, Massachusetts, 1968.

[6] E. Conway and N. Oreskes. *Merchants of Doubt*. Bloomsbury Press, New York, 2010.

[7] S. Crass. Solving the sextic by iteration: A study in complex geometry and dynamics. *Experimental Mathematics*, 8(3):209–240, 1999.

[8] S. Crass. A family of critically finite maps with symmetry. *Publicacions Matemàtiques*, 49(1):127–157, 2005.

[9] S. Crass. Solving the heptic by iteration in two dimensions: Geometry and dynamics under Klein's group of order 168. *Journal of Modern Dynamics*, 1:175–203, 2007.

[10] S. Crass. New light on solving the sextic by iteration: An algorithm using reliable dynamics. *Journal of Modern Dynamics*, 5(2):397–408, 2011.

[11] S. Crass. Criticlly-finite dynamics on the icosahedron. *Symmetry*, 12(1):177, 2020.

[12] S. Crass. www.csulb.edu/~scrass/math.html, 2020.

[13] S. Crass. Completely solving the quintic by iteration. *La Matematica*, forthcoming. arxiv.org/abs/2006.01876.

[14] C. Curtis. *Pioneers of Representation Theory: Frobenius, Burnside, Schur and Brauer*. American Mathematical Society, Providence, 1999.

[15] R. Devaney. *An Introduction to Chaotic Dynamical Systems*. Westview Press, Cambridge, Massachusetts, 2003.

[16] L. Dickson. *Modern Algebriac Theories*. Sanborn, Chicago, 1926.

[17] P. Doyle. Solving equations through the ages, 2020. `math.dartmouth.edu/~doyle/docs/solve/solve.pdf`.

[18] P. Doyle and C. McMullen. Solving the quintic by iteration. *Acta Mathematica*, 163:151–180, 1989.

[19] M. Duchin. Gerrymandering v. geometry. *Scientific American*, 319(5):48–53, 2018.

[20] R. Feynman. *QED: The Strange Theory of Light and Matter*. Princeton University Press, Princeton, 1985.

[21] B. Franklin. *Mr. Franklin: A Selection from His Personal Letters*. Yale University Press, New Haven, 1956.

[22] R. Fricke. *Lehrbuch der Algebra*. Vieweg, Braunschweig, 1926.

[23] M. Henle. *Modern Geometries: Non-Euclidean, Projective, and Discrete Geometry*. Pearson, London, 2nd edition, 2001.

[24] B. Hunt and E. Kostelich. Dynamics 2, 1998.

[25] S. Johnson. *Farsighted: How We Make the Decisions That Matter the Most*. Riverhead Books, New York, 2018.

[26] D. Kahneman and A. Tversky. Choices, values, and frames. In D. Kahneman and A. Tversky, editors, *Choices, Values, and Frames*, pages 1–16. Cambridge University Press, Cambridge, 2000.

[27] I. Kant. *Groundwork of the Metaphysics of Morals*. Cambridge University Press, Cambridge, 2nd edition, 2012.

[28] S. Kantayya. Coded bias. Seventh Empire Media, 2020. Documentary video.

[29] F. Klein. Über eine geometrische repräsentation der resolventen algebraischer gleichungen. *Mathematsche Annalen*, 4:346–358, 1871.

[30] F. Klein. Vergleichende betrachtungen über neuere geometrische forschungen. *Verlag von Andreas Deichert*, 1872. English translation at `arxiv.org/abs/0807.3161`.

[31] F. Klein. Zur Theorie der allgemeinen Gleichungen sechsten und siebenten Grades. *Mathematische Annalen*, 28:499–532, 1886.

[32] F. Klein. *Lectures on the Icosahedron*. Dover Publlications, New York, 1956.

[33] F. Klein. *Elementary Mathematics from an Advanced Standpoint*. Dover Publications, New York, 2004.

[34] M. Kline. *Mathematical Thought from Ancient to Modern Times*. Oxford University Press, Oxford, 1972.

[35] M. Lewis. *The Undoing Project*. W. W. Norton, New York, 2017.

[36] C. McMullen. *Algebra and Dynamics*. Harvard Lecture Notes (unpublished), 2012.

[37] C. McMullen. Julia, 2021. `people.math.harvard.edu/~ctm/programs`.

[38] T. Needham. *Visual Complex Analysis*. Oxford University Press, New York, 1997.

[39] D. Neuenschwander. *Emmy Noether's Wonderful Theorem*. Johns Hopkins Press, Baltimore, 2011.

[40] Plato. *Timaeus*. Harvard University Press, Cambridge, Massachusetts, 1929.

[41] T. Schelling. *Micromotives and Macrobehavior*. Norton Press, New York, 1978.

[42] J. Shurman. *Geometry of the Quintic*. Wiley, New York, 1997.

[43] C. Sunstein and R. Thaler. *Nudge: The Final Edition*. Penguin, New York, 2021.

[44] F. Toscano. *The Secret Formula*. Princeton University Press, Princeton, 2020.

[45] W. Tschirnhaus. A method for removing all intermediate terms from a given equation. *Acta Eruditorum*, pages 204–207, 1683. English translation by R. Green.

[46] K. Ueno. Dynamics of symmetric holomorphic maps on projective spaces. *Publicacions Matemàtiques*, 51:333–344, 2007.

[47] H. Valentiner. De endelige transformations-gruppers theori. *Videnskabernes Selskabs Skrifter: Naturvidenskabelig og Mathematisk Afdeling*, 2, 1889.

[48] A. Wiman. Ueber eine einfache gruppe von 360 ebenen collineationen. *Mathematische Annalen*, 45:531–556, 1895.

Index

Milton Keynes UK
Ingram Content Group UK Ltd.
UKHW020805210124
436410UK00006B/8

9 780367 564933